SUFFERING MADE REAL

SUFFERING MADE REAL

THE UNIVERSITY OF CHICAGO PRESS

American Science
and
the Survivors at
Hiroshima

M. Susan Lindee

Chicago & London

M. Susan Lindee is assistant professor in the Department of the History and Sociology of Science at the University of Pennsylvania.

The University of Chicago Press, Chicago 60637
The University of Chicago Press, Ltd., London
© 1994 by The University of Chicago
All rights reserved. Published 1994
Printed in the United States of America
03 02 01 00 99 98 97 96 95 94   5 4 3 2 1

ISBN (cloth): 0–226–48237–5

♾ The paper used in this publication meets the minimum requirements of the American National Standard for Information Sciences— Permanence of Paper for Printed Library Materials, ANSI Z39.48–1984.

Library of Congress Cataloging-in-Publication Data

Lindee, M. Susan.
    Suffering made real : American science and the survivors at
Hiroshima / M. Susan Lindee.
        p.   cm.
    Includes bibliographical references and index.
    ISBN 0-226-48237-5 (acid-free paper)
    1. Radiation—Physiological effects—Research—Social aspects.
2. Atomic Bomb Casualty Commission.   3. Hiroshima-shi (Japan)—
Bombardment, 1945—History.   4. Atomic bomb victims—Japan—
Hiroshima-shi.   5. Children of atomic bomb victims—Japan—
Hiroshima-shi.   I. Title.
RA1231.R2L497   1994
363.17′99—dc20   94-1832
                                                               CIP

To my parents

Contents

Acknowledgments

I am sure it must be easier to work on historical topics in which the principals have departed this world. However, there are advantages to dealing with events in the relatively recent past, since many of the relevant actors are still able (and willing) to complain about one's analysis. James V. Neel gave me access to his papers, participated in lengthy oral history interviews, and critically read this manuscript. He does not agree with my assessment of the Atomic Bomb Casualty Commission, feeling, as do many other ABCC veterans, that I have overemphasized the impact of political and social concerns on the science of the ABCC.

I do think that Neel and his colleagues struggled heroically to conduct their science in that neutral zone in which language, culture, and history do not exist, that is, in the realm of the idealized Science that they learned in the course of their formal education. My text operates from the assumption that such a neutral zone does not exist, for anyone, at any time. Despite our differences, Neel's input, including his objections and counterarguments, has dramatically improved my work and I thank him for his time and sustained interest.

I must also especially thank William J. Schull, Grant Taylor, and the other ABCC-RERF veterans who spoke with me, corresponded, or read various drafts or portions of drafts. As mentioned above, most of these ABCC veterans are not sympathetic to my portrait of the organization and its history.

Dorothy Nelkin and William B. Provine worked closely with me throughout this project, which began as my doctoral dissertation. Others who have read drafts or chapters and offered helpful suggestions include James F. Crow, Margaret Rossiter, Richard Polenberg, Stephen Hilgartner, Keith Wailoo, Charles Rosenberg, Harry Marks, and Atsushi Akera. Vassiliki B. Smocovitis's reading of a near-final draft helped in more ways than she may realize. John Beatty generously encouraged and aided a junior scholar and I appreciate his help immensely. John's perceptive insights at crucial stages—and his careful, critical readings of two full drafts—helped me around some of my most aggravating problems.

Many archivists provided guidance and assistance, particularly Janice Goldblum at the National Academy of Sciences, who cheerfully helped me find my way through the maze of ABCC records retained there; also, Michael Rhode at the Armed Forces Institute of Pathology, Roger M. Anders at the Department of Energy Archives, Martin Levitt at the American Philosophical Society, and Elizabeth Borst White at the Houston Academy of Medicine/ Texas Medical Center. Other archives that made resources available were the Wilmington College Peace Resource Center in Wilmington, Ohio, the Bentley Historical Collection at the University of Michigan, the Rockefeller Archives Center in Pocantico Hills, New York, and the National Library of Medicine in Bethesda, Maryland.

My work was supported by the National Science Foundation and by the Center for International Studies at Cornell University. In addition, the Department of the History and Sociology of Science at the University of Pennsylvania and the Beckman Center for the History of Chemistry supported my work on the final draft by providing me with a full semester's respite from teaching before I had even begun and, later, a research assistant. Andrew W. Mellon Foundation funding brought me to Penn, where I have enjoyed all the benefits of a lively intellectual community. Special thanks to my colleagues at Penn.

An earlier version of chapter 7 appeared in 1994 in the *Bulletin of the History of Medicine,* vol. 68, no. 3, as "Atonement: Understanding the no-treatment policy of the Atomic Bomb Casualty Commission." Sections of chapter 9, in modified form, appeared in 1992 in the *Journal of the History of*

Biology, vol. 25, no. 2, as "What is a mutation? Identifying heritable change in the offspring of survivors at Hiroshima and Nagasaki." I thank these journals for permission to use this material here.

On the personal level, thanks are due to Lazara DeBrito, Margaret Ferry (and Cheryl and Christine), and Joanne Grier. I am grateful to Sydney Rubin for moral support and trans-Atlantic phone calls. Finally, I wish to thank my husband Brett Skakun, for counsel and courage, and our children Grant and Travis, for delightfully interfering with my work.

 ONE

How
the
ABCC
Began

CHAPTER ONE

The Most Important People Living

here were no battles in Hiroshima or Nagasaki. At 8:15 A.M. on 6 August 1945, Hiroshima was a busy, functioning urban center with a large population and complex support and emergency systems. One minute later, Hiroshima—its central four square miles—was gone.[1] On 9 August, the Urakami Valley of Nagasaki was similarly transformed. For those who sur-

1. The zone of complete destruction in Hiroshima was an area of about four square miles. Most buildings within an area of 9.5 square miles were destroyed or badly damaged. Window glass was shattered up to eight miles beyond the central zone of complete destruction. The area completely burned covered 4.5 square miles. The total area of severe damage was somewhat less in Nagasaki because of the hilly terrain. Complete destruction occurred within 1.5 miles of the hypocenter, with significant destruction to 2.5 miles. (Oughterson and Warren 1956, 15–17, 20).

vived, the unthinkable became the commonplace.[2] They had been eating an orange, working in a garden, or reading a book. Minutes later they wandered, without feeling, past corpses, neighbors trapped in burning mounds of rubble, or children without skin. Hiroshima survivor Wakashi Shigetoshi has written that the suffering he saw around him in the immediate aftermath of the bombing "no longer moved me in the slightest. At that time human beings on the point of death were no longer human: they became mere substance. And the man watching them lost his humanity, and also became but a substance" (Takayama 1973, 63). Robert Jay Lifton, an American psychiatrist who interviewed Hiroshima survivors in 1962, characterized the experience as "psychic closing off," a complete cessation of emotion that occurred "very quickly—sometimes within minutes or even seconds" (Lifton 1968, 31). The novelist Ōe Kenzaburō has described those who survived this experience as "people who, despite all, didn't commit suicide" (Ōe 1981, 76).

On those two August mornings the residents of Hiroshima and Nagasaki endured what many came to fear would be the fate of humanity as a whole. Those who lived through this initiation were the first true residents of the atomic age. They were profoundly changed by the bomb: physically changed by the injuries they sustained, and by their exposure to high levels of ionizing radiation, which could later lead to disease; psychologically changed by the experience itself, by the pain they endured, by the loss of loved ones, by the horrors they witnessed, and by their own survival when so many around them died; and changed by their status as witnesses, a status that shaped how others understood and interpreted them. All conjectures about the nature of the imagined post–nuclear war world must draw on their experiences at Hiroshima and Nagasaki. The survivors have been key figures in postwar narratives contesting the future of the human species.

Many studies, including the scientific research of the Atomic Bomb Casualty Commission (ABCC), the subject of this book, have incorporated a view of the survivors as a scarce and precious intellectual resource. Robert Holmes,

---

2. The extensive survivor literature includes memoirs, poetry and fiction. See Kyoko Selden and Mark Selden, *The Atomic Bomb: Voices from Hiroshima and Nagasaki* (1989). Also see Agawa Hiroyuki, *Devil's Heritage* (1957); Ibuse Masuji, *Black Rain* (1969); Hachiya Michihiko, *Hiroshima Diary: The Journal of a Japanese Physician, August 6–September 30, 1945* (1955); and Ōe Kenzaburō, *Hiroshima Notes* (1981). For theater, see David Goodman, ed., *After Apocalypse: Four Japanese Plays of Hiroshima and Nagasaki* (1985). I found the survivor paintings and drawings, published as *Unforgettable Fire: Pictures Drawn by Atomic Bomb Survivors,* to be particularly striking artistic translations of an incomprehensible human experience (Japanese Broadcasting Corporation 1977). Nagai Takashi's psychological portrait, *We of Nagasaki: The Story of Survivors in an Atomic Wasteland,* includes several terrifying first-person accounts with a strangely flat emotional tone (1958).

who served as director of the ABCC from 1954 to 1957, told a journalist in 1957, "These are the most important people living" (Trumbull 1957, 132–33). More recently, a Japanese publication on the bombings expressed a similar idea: "Not often are a few chosen people challenged to fulfill so crucial a role in human history" (Committee for Compilation 1981, 500).

HIBAKUSHA

The Japanese call those who lived through the atomic bombings *hibakusha,* roughly, "survivor" or "exposed one." Some observers count as *hibakusha* only those who were exposed to the bombs' direct effects: blast, fire, debris, radiation. Others include those who were not in the cities at the time but who lost family members and property in the blasts. Official Japanese registries in the 1980s recognized 368,259 survivors. Of these the vast majority—71 percent in Hiroshima and 69 percent in Nagasaki—suffered symptoms of acute radiation sickness in the days immediately after the bombings. In-air doses of gamma rays at ground zero in Hiroshima were estimated at 10,300 rads and in Nagasaki at 25,100 rads, and varying degrees of radiation sickness occurred in those exposed within 5 kilometers (3 miles) of ground zero (Committee for Compilation 1985, 106).

To understand the magnitude of these exposure levels, it may be useful to compare them to the exposures believed to have occurred in the 1986 nuclear power plant accident at Chernobyl, in the former Soviet Union. Symptoms of radiation sickness begin at about 100 rads (1 Gray). Exposures at Chernobyl are reported to have peaked at about 500 rads, and only three persons who survived the accident are publicly recognized as having been exposed to levels this high. Approximately four thousand people may have received radiation dosages as high as 200 rads, however, and an additional fifty thousand probably received doses of 50 rads. Official casualty reports at Chernobyl list 31 fatalities and 238 persons suffering "acute radiation sickness," but these numbers are probably far too low. While an additional six hundred thousand persons are classified as having been "significantly exposed" to radiation at Chernobyl, the precise meaning of this designation is unclear (Medvedev 1990, 129–36).

Other populations exposed to high levels of ionizing radiation have included the eighty-two inhabitants of the Rongelap Atoll and the twenty-three Japanese fisherman aboard the *Lucky Dragon,* both groups victims of fallout from the 1954 hydrogen bomb tests at Bikini atoll; early radiologists, who absorbed relatively large cumulative doses; the New Jersey radium dial painters, who in the 1910s and 1920s ingested radioactive materials in the workplace; uranium miners; and patients treated with radiation for such conditions as spondylitis and infertility (Caufield 1989, 30–37, 140–47; Lapp 1958; Gofman 1981). Of these populations, only the Rongelap islanders and the Japanese fishermen received whole body irradiation in a large single dose, the type of exposure that would be expected in the event of future nuclear war. The atomic bomb survivors have therefore had a special place in the study of radiation effects.

For scientific researchers their primary disadvantage as a study population was that the amount of radiation each absorbed was unknown, particularly in the immediate postwar period when even the total amount and type of radiation released by the atomic bombs was a U.S. military secret. Estimating dosage is a problem for scientists studying any exposed population, since human beings cannot ethically be experimentally irradiated under controlled conditions (though recent news reports suggest that the AEC was not constrained by ethical concerns from conducting such experiments [Healy 1993]). But in cases of medical irradiation, for example, a patient's record can at least suggest a limited range of possible exposure levels. Dosimetry for the bomb survivors has been more difficult. Their levels of exposure depended on atmospheric humidity, on their exact body position at the moment of detonation, on whether they were shielded in any way from the explosion and on the properties of the shielding material, on their activities immediately after the explosion, and on whether they ingested radioactive materials. American and Japanese scientists worked for more than forty years on the problem of dosimetry for the survivors and aspects of these calculations remain controversial (Marshall 1992). Despite this, scientists generally concluded that they had to study the survivors. Their research was undertaken with the hope that there would never be another population similarly exposed and with the expectation that there would be.

Identifying all survivors has also proven difficult. The process of reconstructing what happened to people in Hiroshima and Nagasaki was complicated by the chaos that the bombs produced. The standard systems of accounting for people—hospitals, police records, death and birth records—were not functioning in the months after the bombings. Assessments of how many people were killed, and even of the number of survivors, remain inconclusive. Current Japanese estimates state that 90,000 to 120,000 persons died in Hiroshima in the first four months after the bombings. Others estimate as many as 140,000 initial deaths (Selden and Selden 1989, xxi). In Nagasaki, Japanese authorities estimate the number killed in this critical stage at 60,000 to 70,000 (Committee for Compilation 1985, 18–21, 46–48).[3]

3. In contrast, the firebombing of Tokyo on 9 and 10 March 1945 claimed one hundred thousand lives (Selden and Selden 1989, xxii).

These counts are imprecise for a number of reasons. The populations of both cities were in flux in response to wartime needs. Residents of Hiroshima expected their city to be bombed—it was an important military supply center—and many parents sent their children to live with rural relatives. At the same time, many residents from the surrounding areas were called into the city on an irregular basis to serve in work crews, building fire trenches and doing other civil defense work. After the bombings, it was impossible to determine exactly how many people were in the affected areas at the moment the bombs were dropped; the two cities were places of anonymous, uncountable death. Many of the dead were buried under rubble, some of which was not cleared for years. There were no bodies left to count near the hypocenter: the heat and energy literally vaporized the closest persons. And many bodies were swept out to sea with the tides, after dying burn victims sought relief in Hiroshima's numerous rivers.[4] In the weeks after the bombings, relief workers buried thousands of corpses in mass graves as a public health measure, and while they counted the bodies, they often could not identify them (Committee for Compilation 1985, 18–19, 31–33).

Counting the survivors should have been easier than counting the dead, yet this too proved difficult. Except in the case of those seriously injured and still hospitalized long after the bombings, the enumeration of survivors depended almost entirely on voluntary self-identification. Many survivors had substantive reasons for keeping their experiences to themselves. There was a stigma associated with having been exposed to the bomb's radiation. Many believed that survivors would be unable to have healthy children due to radiation-induced heritable mutations, and in a social world where marriages were commonly arranged, some families concealed exposure in an effort to secure desirable matches for their children.[5] In addition, until 1957 survivors qualified for no special assistance or medical care and therefore had little in-

---

4. For several accounts of the scenes at Hiroshima's rivers, see the first-person essays in Takayama Hitoshi, ed., *Hiroshima in Memoriam and Today,* particularly, Wakaki Shigetoshi's "I Hate Hiroshima" (Takayama 1973, 55–114, esp. 60–63).

5. Ibuse Masuji's novel *Black Rain* is the story of a family unable to arrange a marriage for a daughter exposed to the atomic bomb (1969); also see extensive discussion in Robert Jay Lifton's *Death in Life* (1968, 543–55). A survivor who chose to publish only under his initials, N. E., wrote as follows: "I have been using my A-bomb survivor's medical record book since 1958 for my medical treatment, but my wife and son do not have the book. They were within two kilometers of the hypocenter when the bomb fell, so they are, of course, entitled to have the book issued. However, we made a false statement saying that they had been evacuated to Mt. Iwaya by the time the bomb fell. We had to do so because we knew there was discrimination against A-bomb survivors" (Kubo 1990, 108–9). See also *Henken to sabetsu* (1971).

centive to come forward and register. The number of persons who identified themselves as survivors increased steadily after a Japanese law established free medical care for bomb survivors.[6]

Current estimates, based on census surveys begun in 1950 and on continuing Japanese surveys, place the number of survivors at 368,259, as many as eighteen of whom may have lived through both bombings.[7] This estimate of survivors includes some "secondary" victims, that is, rescue workers and others who came into the bombed areas shortly after the bombings (under current Japanese law, within the first three days) and were therefore exposed to residual radiation (though the presence of such radiation remains a point of contention). Some estimates also include "tertiary" victims, people who were not themselves physically present in Hiroshima or Nagasaki but who lost family or property. The number of survivors calculated also depends on the length of survival required for inclusion. "Survivors" could include those who were not immediately killed by the blast but who died hours or days or weeks later. Most estimates count as survivors those who were still alive three or four months after the bombings (Committee for Compilation 1985, 145–46).

Hundreds of personal testimonials by atomic bomb survivors have appeared since 1945—rates per year have fluctuated, from a low of one such publication in 1947, to 228 in 1971.[8] Their stories have served many ends. For psychologists, the tales have documented the effects of trauma. For peace activists, the survivors' accounts have served to promote their cause by demonstrating the tragedy of war. For medical researchers, the survivors' experiences,

---

6. This was the A-Bomb Victims Medical Care Law (1957) (Committee for Compilation 1985, 145).

7. Robert Trumbull interviewed nine such "double survivors," persons who were present in Hiroshima during the first attack, then traveled for various reasons to Nagasaki and were there during the second attack three days later. One man was bringing the bones of his wife, killed in the Hiroshima attack, to her parents' home. Another was just finishing a three-month assignment in a Hiroshima shipyard and was scheduled to return to his family in Nagasaki on 7 August. According to records of the Atomic Bomb Casualty Commission and some Japanese official records, Trumbull states, there were nine other persons who also survived both bombings. See Trumbull's *Nine Who Survived Hiroshima and Nagasaki* (1957).

8. Some years saw a flurry of activity: 145 testimonials were published in 1951, 284 in 1954 and 1955 (at the height of the public debate over atmospheric weapons testing), and 205 in 1965 (perhaps related to the Vietnam war). Also see, for example, the pamphlet prepared by the A-Bomb Victims Association (Atomic Bomb Victims' Written Notes Editorial Committee 1953). See the discussion of the peace movement in *Hiroshima and Nagasaki: The Physical, Medical and Social Effects of the Atomic Bombings* (Committee for Compilation 1981, 575–93; a table showing rates of publication of survivor literature appears on page 586). See also, Hachiya's *Hiroshima Diary* (1955) and Arata Osada's *Children of the A Bomb: The Testament of the Boys and Girls of Hiroshima* (1959).

and their bodies, have promised to reveal the risks posed by radiation. For others, including many American political leaders, the survivors have been necessary Cold War martyrs whose suffering demonstrated the power of the West.[9]

Some survivors chose to take an active public role in protesting war and promoting peace; others rejected political activism. Survivors' published testimonials are often antiwar, either explicitly or implicitly, and some survivors have testified at peace rallies—for example, Chieko Watanabe at the Second World Conference against Atomic and Hydrogen Bombs in 1956. Survivors have also put together their own literature to protest nuclear weapons, such a Arata Osada's collection of children's testimonials, *Children of the A-Bomb* (1959). At the same time, some survivors have expressed feelings of being exploited by the Japanese antiwar movement. These survivors have seen themselves as a "core of authenticity" taken advantage of by self-serving forces. Some have accused the peace movement of "selling the disaster" and "using the A-bomb as their flag."[10]

Survivors' victimization took the concrete form of the failure of the Japanese government to assist in their recovery. The government failed to meet their acute needs for medical care and social support at least until 1957, when the first version of the A-Bomb Victims Medical Care Law took effect.[11] Survivors have also been socially ostracized and rejected as marriage partners out of fear

9. Psychological studies include Irving L. Janis, *Psychological Aspects of Vulnerability to Atomic Bomb Attacks* (1949) and *Air War and Emotional Stress* (1951); Lifton's *Death in Life* 1968); T. Misao's article "Characteristics in Abnormalities Observed in Atom-Bombed Survivors" (1961); and the report *The Effects of Strategic Bombing on Japanese Morale* by the U.S. Strategic Bombing Survey (1947). The problem of the peace literature and the peace movement is explored by Lifton (1968, 270–87). See also George O. Totten and T. Kawakami, "Gensuikyō and the Peace Movement in Japan" (1964). The survivors' construction as medical research subjects is the focus of this book. On the interpretation of the survivors as Cold War sacrifices, see Gar Alperovitz, *Atomic Diplomacy, Hiroshima and Potsdam: The Use of the Atomic Bomb and the American Confrontation with Soviet Power* (1965), and P. M. S. Blackett, *Fear, War and the Bomb: Military and Political Consequences of Atomic Energy* (1949).

10. Lifton explores survivors' feelings about the peace movement in perceptive detail (Lifton 1968, 290–305).

11. The first version provided for health examinations and limited governmental payment for some medical treatment of survivors. Revisions in 1960 and 1962 expanded the scope of the law, increasing the number of persons included and the range of medical treatment to be provided. The 1962 revision included "early entrants," that is, anyone who had entered Hiroshima or Nagasaki within two weeks of the bombings, as survivors. In 1965, early entrants who qualified for free medical care were redefined as those who had come into either of the two cities in the first three days after the bombings. A report by the Committee for the Compilation of Materials on Damage Caused by the Atomic Bombs in Hiroshima and Nagasaki describes the history of the passage of this law and its various amendments (Committee for Compilation 1981, 542–51).

that they would be unable to have healthy children.[12] Many suffered economic hardship as a consequence of their physical disabilities, their inability to work, and widespread discrimination against bomb survivors (Committee for Compilation 1981, 431–33). Finally, the survivors have been the subject of extensive scientific research, both Japanese and American, almost continuously since 1945, and while this research has provided the survivors with valuable information about their medical status, it has also in some ways victimized them.[13]

I begin my exploration of the Atomic Bomb Casualty Commission with the survivors because of the important role they play in the story that follows. I argue that the survivors themselves shaped the scientific work of the ABCC. Many of the policies and decisions of ABCC administrators and scientists can be understood as reactions to the special cultural and political status—the historical place—of the ABCC's unique research subjects.

WHY STUDY THE SURVIVORS?

The bomb's peculiar form of terror, ionizing radiation, was known to cause biological changes in both humans and experimental organisms. This made the long-term medical studies of the atomic bomb's victims a high priority, for unlike victims of conventional bombing, their bodies' responses to the effects of the bombs might take decades to appear.

The atomic bomb's physical destruction—caused by blast and fire—was in essence no different from that caused by conventional weapons. While U.S. military planners were interested in assessing how much damage conventional bombing inflicted, they were only marginally interested in the medical effects of such bombing.[14] Military studies of bomb victims in Germany were carried out during the Allied occupation, but these studies simply compiled historical data for the years 1938 to 1944, detailing the causes of death in heavily bombed

12. A 1965 survey of atomic bomb survivors asked respondents if they had experienced "adverse discrimination" in marriage. Of those in the unmarried 35-to-39 age group, 11.4 percent reported such discrimination; the overall rate was 2.6 percent (Committee for Compilation 1981, 427). Also see note 5 above.

13. The victimization of the survivors by the Japanese government is explored in *Hiroshima and Nagasaki: The Physical, Medical and Social Effects of the Atomic Bombings* (Committee for Compilation 1981, 339–40). The same report discusses social problems encountered by the survivors, on pages 420–26.

14. American planners deliberately spared possible target cities—Kyoto, Hiroshima, Kokura, Niigata, and Nagasaki—from conventional bombing, so that the full physical impact of the atomic bomb could be assessed (Selden and Selden 1989, xvii; Rhodes 1986, 632, 639–41).

areas. They did not involve long-term epidemiological studies of survivors; military physicians were instead concerned with how effective Allied bombing raids had been in disrupting human activities.[15]

The long-term studies of the atomic bomb survivors differed from these other studies in that they involved a fundamental biological problem with ramifications far beyond military strategy. The survivors had been exposed to a form of energy that was widely used in medical therapy and known to cause heritable mutations and other cellular changes in experimental organisms.[16] A formal system setting limits on human exposure to radiation was already in place in Britain and the United States (Whittemore 1986). The problems to be explored at Hiroshima and Nagasaki involved a recognized but poorly understood hazard to human health with long-term effects that were not immediately apparent in the survivors. These invisible effects could be revealed only through epidemiological research and statistical analysis.

Radiation released by the bombs killed the residents of Hiroshima and Nagasaki in ways that initially baffled Japanese physicians. Many survivors who seemed to come through the blast unhurt began, within minutes, hours, or days, to manifest the symptoms of acute radiation sickness. Sudden, severe nausea and diarrhea were the first signs, followed later by subcutaneous bleeding and gingivitis (inflammation of the gums). Many of these seemingly unhurt survivors died in the days and weeks after the bombings, of an illness Australian journalist Wilfred Burchett called "atomic plague."

Burchett, who arrived in Hiroshima on 4 September, was the first to report this condition in the international press, but his story was dismissed by Manhattan District (U.S. Army Corps of Engineers) scientists as Japanese propaganda.[17] This dismissal was not entirely political: American authorities were genuinely skeptical of his story, because Manhattan Project scientists at Los Alamos felt it could not possibly be correct. They did not deny that the bomb had released radiation; that had been expected. Rather, as Peter Wyden's 1984 study suggests, they did not believe that anyone close enough to the hypocenter

15. Topics explored by the team of military physicians included the nature of air raid casualties (carbon monoxide poisoning, dust inhalation, heat, blast), communicable diseases, effects of bombing on industrial workers, public health problems, and effects of bombing on German health care systems (U.S. Strategic Bombing Survey 1945).

16. Important publications prior to 1945 dealing with radiation effects include P. S. Henshaw, "Experimental Roentgen Injury" (1944a); Shields Warren, "Effects of Radiation on Normal Tissues" (1942b); and L. C. Fogg and S. Warren, "Some Cytologic Effects of Therapeutic Irradiation" (1941). See also G. L. Clark, "A 1936 Survey of the Biological Effects of X-Irradiation" (1936).

17. See Wilfred Burchett's account of his experiences in September 1945 in Japan, *Shadows of Hiroshima* (1983, 15–24, 34–39).

to be severely irradiated could survive the blast. They thought that "any person with radiation damage would have been killed with a brick first." [18] But many did survive in locations relatively close to the hypocenter. Approximately half of those within three-quarters of a mile of the hypocenter at Hiroshima survived the initial blast, and at least 131 persons who had been within two kilometers (1.2 miles) of the hypocenter were still alive in 1953. [19]

When Burchett's story was published, the atomic bomb survivors at Hiroshima and Nagasaki were already the focus of medical studies by Japanese investigators. Within weeks, these Japanese physicians and scientists were joined by three American military teams. Almost two years later, a permanent American agency to investigate the long-term effects of radiation on the survivors was established at Hiroshima and Nagasaki. This was the Atomic Bomb Casualty Commission.

My historical account focuses on the scientific work carried out on the survivors by the ABCC in the first decade after the bombings. I explore the processes (administrative, intellectual, personal) through which the unprecedented events in Hiroshima and Nagasaki became elements in the construction of data. I deal with the ways in which data were collected, analyzed, and publicly articulated, showing how pieces of the social world and the natural world interacted. My analysis explores how each step of removal and abstraction— each step of scientific distancing—was constrained and pressured by cultural setting. This is a story about turning life into science and about the transformation of incomprehensible human suffering. I ask why policies were important—why they were sustained—and how particular choices were made about managing the survivors (treatment, public relations), the institution (funding, affiliation), and the data (what counts as a control? what problems justify dumping data?). I show how a terrifying human event, the explosion of the atomic bomb, discouraged some choices and encouraged others.

My subject is an agency that has been the focus of hostile published ac-

18. The quote is from Norman Ramsey, who was chief of the Delivery Group at Los Alamos, in Peter Wyden's *Day One: Before Hiroshima and After* (1984, 16). Wyden suggests that Manhattan Project scientists have persisted in this disbelief. Joseph O. Hirschfelder, who was responsible at Los Alamos for predicting the effects of the bomb, told an academic meeting in 1980 that he did not believe there were more than one or two thousand people who received radiation burns in the bombings but were not killed. Yet even Stuart Finch, onetime director of the Radiation Effects Research Foundation (successor agency to the ABCC, which was renamed and reorganized in 1975) has estimated that twenty thousand people sustained radiation injuries but survived the bombings. Wyden suggests that even this number is far too low (Wyden 1984, 325–26).

19. Most of those who had been near the hypocenter died in the days or weeks following the bombing. The majority of the long-term survivors had been at more distant locations (Committee for Compilation 1985, 147).

counts. I want to emphasize that this book is neither an apology for the ABCC nor an attack on it.[20] The ABCC was a complicated institution in a complicated situation. The Americans were studying a population devastated by a terrible new American weapon, in a country under American military occupation. Their research required body fluids, information about abnormal births, and survivors' bodies for autopsy. The sensitive nature of this material heightened emotions that may have been less volatile when played out over labor policy or censorship issues. The ABCC's policies and decisions were in many ways no more insensitive to Japanese needs than any other Allied policies, but they dealt with issues of extreme personal urgency.[21] Relatively minor transgressions by the ABCC could provoke intense community responses, partly because they reflected the power relationship in symbolically important ways. Many attacks on the ABCC dealt more with its cultural meaning than with its actual practices: tensions in Japanese-American relations during and immediately after the Allied military occupation were projected vividly onto the day-to-day operations of the ABCC.

I should also state that biological reality is not the central question in this book. The scientists involved have engaged in extensive debates about biological reality for decades. They continue to disagree about dosimetry, extrapolation from experimental organisms, risks of low-level exposure, and many other questions. Radiation biology, like most scientific fields, is contentious. My goal is not to resolve these debates; I could not possibly do so. I have instead documented the social and cultural realities that produced one important element in these debates, namely, the data on the genetic effects of radiation on the offspring of survivors of the bombings. I have not asked *whether* heritable mutations occurred in the populations at Hiroshima and Nagasaki; I assume, as do most geneticists, that they did. I have asked, rather, how was the problem

---

20. For hostile accounts, see survivor testimonials in Lifton's *Death in Life* (1968, 343–65), Robert Jungk's *Children of the Ashes* (1961, 239–45), and various Japanese critiques of the ABCC, including *Gensuibaku higai hakusho: Kakusareta shinjitsu.* [White paper on hydrogen and atomic bomb victims: The hidden truth] (1961); *Gembaku, Gohyakunin no shōgen: Hibakusha tsuiseki chōsa repōto* [Five hundred a-bomb testimonies: Report of a hibakusha survey] (1967); *Hibaku nisei: Sono katararenakatta hibi to asu* [Hibaku nisei: Their untold past and their futures] (1972); *Kagaisha e no ikari: ABCC wa nani o shita ka* [Anger toward the agressor: How the ABCC treated us] (1966); and the novel by Kajiyama, *Jikken Toshi* [Experimental City] (1954). See also periodicals cited extensively in my chapter 8, on public relations.

21. One ABCC staff member observed that the people in Hiroshima were convinced "that ABCC does nothing but experiment on patients and has as its ultimate goal an autopsy on each patient." Clarence M. Tinsley to Robert H. Holmes, 18 November 1955, box 32, Atomic Bomb Casualty Commission—National Academy of Sciences—National Research Council, Thomas Francis papers, BHL.

of heritable mutations in the survivors constructed and understood? What forces were at work in the production of the published scientific results? Why did some problems take precedence over others? And why did publication take the form it did?

Following Helen Longino, I argue that it is not satisfactory to look to methodological inadequacies as the means by which values and cultural beliefs make their way into science. To search for poor methodology is to suggest that a proper methodology could produce value-free science. Nor is it satisfactory to suggest that scientific methodology is completely without power to provide access to biological truth. Longino argues that all data are evidence only in a context of background assumptions about the nature of the problem and the nature of the natural world. These assumptions provide a connection between the data and the processes described by the hypothesis being tested. If science were produced by a single person, the influence of background assumptions would suggest that scientific knowledge is irrational or relativistic. But if the background assumptions emerge from the "transformative interrogation of the scientific community," this minimizes the impact of individual idiosyncracies. It also means, however, that community values, assumptions, and beliefs can be incorporated into the theories and propositions expressing scientific knowledge at any given time (Longino 1990, 215–17). Science is a cultural product and science is constrained by nature. These are complementary, not contradictory, statements.

The choices made by Japanese investigators—their assessment of the important questions and the most pressing problems—were often different from those made by American investigators. Some Americans have interpreted these differences as evidence that the Japanese were engaged in "bad science." I would suggest that they were rather manifestations of the differing cultural frameworks within which Japanese and American scientists decided what information mattered, or, for that matter, what counted as good science.

The values that shaped the work of the ABCC were widely shared in the American scientific community in this period. Some of the scientific debates about "genetic load" and species decline were explicit reactions to the development of atomic weapons. But more fundamentally, the basic definition of what mattered—what questions had to be answered—was a product of the political realities of the Cold War.

After it had been used, after it had ended the Pacific war, the atomic bomb posed problems for those who had created it and those who had chosen to use it, that is, for American physicists and American political and diplomatic leaders. I want to suggest that it also posed problems for Americans in general, and indeed for people of all nations. The bomb was a frightening manifestation of technological evil, so terrible that it needed to be reformed, transformed, managed, or turned into the vehicle of a promising future. It was necessary, somehow, to redeem the bomb.

For many, including many physicists, atomic energy was the agent of redemption, the means by which the weapon of ultimate destruction could become a source of productive energy and prosperity. The nuclear power industry is the physical consequence of this powerful need.[22] For others, including American diplomatic experts and political leaders, the bomb was to be salvaged as guarantor of peace. If the weapon was terrible, it was also the key to preventing future war.[23] My study focuses on the biological interpretation of the bomb by American scientists. The biologists' approach to rehabilitating the bomb did not parallel that of the physicists and politicians; few proclaimed that radiation improved the residents of Hiroshima and Nagasaki by inducing positive mutations (though such an idea occasionally appeared in the popular press). Rather, geneticists involved in studying the survivors had the relatively unsatisfying task of putting limits on the damage that the bomb's radiation could do. They could only say, "Here is as far as it goes. Here is as bad as it gets." Their work could perhaps make a cataclysmic event manageable but could never make it beneficent. Politicians salvaged the bomb by interpreting it as a diplomatic weapon to prevent future war, physicists by developing it as a source of energy, but biologists could only say that its radiation caused no more genetic damage than might be expected in mice.

I begin in chapters 2 and 3 with the story of the Joint Commission and the initial field survey by the new ABCC in 1946. In chapter 4, I examine the scientific justification for the genetics project, which was the ABCC's most visible and highly promoted study. In chapter 5, I explore the day-to-day operations of the genetics project and the relationship between the American staff of the ABCC and the Japanese midwives, mothers, and employees who consti-

---

22. David Lilienthal explores this question in his *Change, Hope and the Bomb* (1963). He writes, "The basic cause [of the inflated hopes for atomic energy] was a conviction, and one that I shared fully and tried to inculcate in others, that somehow or other the discovery that had produced so terrible a weapon had to have an important peaceful use. Such a sentiment is far from ignoble. . . . Everyone . . . wanted to establish that there is a beneficial use of this great discovery. We were grimly determined to prove that this discovery was not just a weapon. This led, perhaps, to wishful thinking" (109–10).

23. The bomb's limitations as a diplomatic tool were not immediately realized. Martin Sherwin's *A World Destroyed: The Atomic Bomb and the Grand Alliance* explores the ways that both scientists and policymakers interpreted the bomb as a possible means of controlling postwar world affairs (1975). The idea that the bomb was conceived and used primarily as a diplomatic tool (to frighten the Soviet Union) has been extensively discussed with various conclusions. See P. M. S. Blackett, *Fear, War and the Bomb* (1949); Gar Alperovitz, *Atomic Diplomacy* (1965); and Herbert Feis, *The Atomic Bomb and the End of World War II* (1966).

tuted the data collection system. In chapter 6, I show how the ABCC's over-seers in Washington—particularly at the Atomic Energy Commission—assessed the value of the ABCC, and how they reacted to the outbreak of the Korean war by cutting the study's funding. I then analyze, in chapter 7, one of the ABCC's most controversial policies, the policy prohibiting medical treatment of the survivors who were the subject of ABCC biomedical research. In chapter 8 I explore the ABCC's public image in both Japan and the United States, showing how staff members reacted to both negative and positive publicity and how the organization managed its public relations through a program of exit interviews and rumor control.

In chapters 9, 10, and 11 I focus on the genetics data and its scientific analysis. Chapter 9 assesses the meaning in practice of the term "mutation," suggesting that the working definition of mutation in the ABCC genetics study reflected public concerns about the genetic effects of radiation. Chapter 10 focuses on the first analysis of the genetics data in 1952 and 1953, an analysis that led the National Academy of Sciences and the Atomic Energy Commission to agree to stop collecting data on newborns in Hiroshima and Nagasaki in early 1954. In chapter 11, I show how the two most important public texts of the ABCC genetics project, the 1953 preliminary essay in *Science* and the 1956 monograph by James V. Neel and William J. Schull, were developed and constructed. Chapter 12 brings the story up to date, with a summary of the ABCC story since 1956, briefly examining the Frances Committee report, the directorship of George Darling, and the 1975 renaming and reorganization of the ABCC as the Radiation Effects Research Foundation. In my conclusion, chapter 13, I explore the phenomenon of distancing that has characterized not only the work of the ABCC, but other responses to the terrifying events in Hiroshima and Nagasaki in August 1945.

CHAPTER TWO

Colonial Science

The Atomic Bomb Casualty Commission was an extension of work undertaken by an American military "joint commission" sent to assess the immediate biological damage done by the bomb in the fall of 1945. The American efforts also drew on the work of Japanese scientists already studying the survivors in September 1945. The ABCC therefore had roots in both the American military studies and the Japanese studies carried out in 1945 and 1946.

While American and Japanese scientists unquestionably worked together to some degree, tensions on both sides affected the way that collaboration was perceived. Many Japanese saw the American scientists as responsible, by virtue of their nationality, for the suffering in the two cities. At the same time, many Americans saw the Japanese as incapable of approaching the problem "objectively." An exaggerated version of the American narrative would be that American scientists arrived in Japan in 1945 and corrected the well-meaning but misguided Japanese scientists, who did not understand the scientific necessity of strictly comparable controls in epidemiological studies.[1] A different story—of American scientists arriving and raiding Japanese studies for information and, more importantly, tissue samples—was put forth by Japanese critics.[2] In this narrative, American scientists appear as unethical, "confiscating" Japanese data and tissue samples and publishing conclusions drawn from this data without properly crediting the Japanese scientists involved.[3]

While neither construction was entirely accurate, each had some basis in fact. The Americans did borrow tissue samples, data, and data collection systems from Japanese scientists, and not all contributions were credited, nor were

1. See, for example, Shields Warren, "Preliminary Report of Investigation of Atomic Bomb Casualties in Hiroshima and Nagasaki, April, May and June 1947," Committee on Atomic Casualties, Organization, NAS. Also, at the eighth meeting of the Committee on Atomic Casualties, held on 20 April 1948, it was said, "The Japanese have made some observations which should be considered. In Nagasaki the city undertook a survey and turned over 6,298 questionnaires to the Atomic Bomb Casualty Commission. The tabulations show how poorly the study was performed and some of the pitfalls to be avoided." ("Minutes" 1948, 80). It is interesting to note that Samuel Coleman cites Harry Kelly, scientific liaison with SCAP and an MIT physicist, as observing in 1946 that in Japan "pure science is extremely advanced." This quotation was drawn from an account by a Japanese physicist, however, rather than from Kelly's own reports (Coleman 1990, 234).

2. In a 1961 white paper published in Hiroshima, a similar charge was made explicitly against Yale pathologist Averill Liebow. Liebow obtained specimens from a Japanese pathologist working at Ujina, near Hiroshima. The white paper said these specimens were "callously confiscated" and then used in Liebow's own publications without any mention of their source. The same document also stated that the "U.S. government wanted to use Japanese scientists as far as possible in collecting data in the bombed areas, which were dangerous to enter on account of residual radiation." (*Gensuibaku higai hakusho* 1961, 18). The problem of continued victimization is provocatively discussed in Lifton (1968) in the context of the survivors themselves. I have extended this to the larger community, particularly to the physicians and scientists who felt they had a prior claim to information and material obtained from the survivors as research subjects. The American actions were therefore a usurpation of a set of data to which the Japanese felt they had legitimate prior claim.

3. See, for example, the translation of Imabori Seiji's *The Era of A- and H-Bombs*. The translation is filed in drawer 2, ABCC Press mid-1957–1964, NAS. Also see (*Gensuibaku higai hakusho* 1961). Nishimori Issei, a physician who was a medical student in Nagasaki at the time of the bombing, expressed the idea that Japanese body parts were stolen in an interview with Monica Braw. He said: "They took all the autopsy material which we had collected and sent it to America. Had even half of it been left, we pathologists could have done research on the effects of the atomic bomb on human beings. As it was, there was no autopsy material [in] the important period of the bombing and the time immediately following it. The material was not returned to us until after thirty years, and then only when we had repeatedly asked for it. At that time, of course, it was already history" (Braw 1986, 12).

all samples returned.[4] The Japanese did have different medical practices, different scientific expectations, and a different idea of what could qualify as an adequate control population for the heavily exposed survivors.[5]

American administrators and scientists commonly praised Japanese contributions to the biological studies in public forums, while observing in internal correspondence and reports that the Japanese lacked the scientific knowledge, the objectivity, or the financial resources necessary to carry out independent medical studies of the survivors.[6] Those involved in planning and organizing

4. Eventually the tissue samples became a diplomatic problem. In 1972 and 1973, much of this biological material was returned to the government of Japan. For records of this return, see boxes 12–15, papers of the Joint Army-Navy Commission, OHA.

5. The different perception regarding a proper control population was partly a consequence of the general Japanese conviction (which is still held) that residual radiation at the two bombing sites was significant enough to have biological effects on early entrants, such as rescue workers. Radiation sickness has been reported in the Japanese literature in many persons who entered the bombed cities within one hundred hours after the bombings. The ABCC used "lightly exposed" survivors as a control for some studies, and interpreted early entrants as "unexposed." This is a point on which Japanese assessments often disagreed with American assessments. The Hiroshima Peace Memorial Museum's literature states that "anyone who entered the area within 1 kilometer of the hypocenter to aid victims or search for relatives within 100 hours of the bombing was considerably affected by exposure to gamma rays." This is part of a statement included in the map given to visitors when I was there in 1988. A typical Japanese criticism of the ABCC's use of "lightly exposed" survivors as controls was expressed by a faculty member at Hiroshima University Medical School in 1966: "Comparison is altogether meaningless when A-bomb survivors comprise one-fifth of the control group. Naturally there is no difference when the exposed and the exposed are compared." Sugihara Yoshio, Hiroshima University Medical School, quoted in "What Has ABCC Done to Date," translation of article in *Asahi Weekly,* 12 August 1966, drawer 2, ABCC Press 1965–1966, NAS.

6. For praise, see virtually any published account of either the Joint Commission or the ABCC. According to Averill Liebow's published diary of the early Joint Commission work in Japan, "as soon as the Japanese discovered that the word 'joint' meant both nations were to labor shoulder to shoulder to meet the formidable task of the medical study, and that they would be treated not as enemies vanquished but rather as colleagues and equals, they gave unstintingly of their thought and labor." Liebow was a Yale pathologist who served at the 39th General Hospital on the island of Saipan, 1942–1945. He joined Ashley Oughterson's Army team in the fall of 1945. His diary, taken in shorthand each night he was in Japan, provides a detailed account of the day-to-day operations of the Joint Commission. Herman Tarnower (later co-author of *The Complete Scarsdale Medical Diet Plus Dr. Tarnower's Lifetime Keep-Slim Program* [1978]), was among the team that traveled to Tokyo with Liebow in September 1945. See "Encounter with Disaster: A Medical Diary of Hiroshima, 1945]" (Liebow 1945). For censure, see for example Fred M. Snell, "Memorandum for the Record," 25 November 1947, JVN; James K. Scott to Charles L. Dunham, 14 October 1954, drawer 4, Atomic Energy Commission 1951–61, NAS; and John C. Bugher, "Supplement to Report of the Atomic Bomb Casualty Commission Field Operations, 27 March to 7 May 1951" and "Report," drawer 28, Dr. John C. Bugher (AEC), Report on ABCC Visit—March–May 1951, NAS, 28.

the American studies of the survivors believed that the Japanese could not independently conduct a scientific study of the survivors. This belief—based on their assessments of both Japanese science and the Japanese economy—justified American intervention. Without it, the American studies would not have made sense. At the same time, Japanese scientists were wary of American intentions and skeptical of individual American scientists. They were not prepared to cede the field to the Americans, and they continued their own studies of the survivors despite the American presence in Hiroshima and Nagasaki.

The common American claim during the Occupation that the scientific work was a "joint venture" was nonetheless more descriptive of actual practice than the strategists using the phrase may have realized. The American work depended on Japanese workers and subjects, reflected Japanese culture, and was shaped by the social climate of postwar Japan.

The American Occupation operated for the most part by issuing directives to preexisting Japanese institutions, which then carried them out. General Douglas MacArthur, who was in charge of the military occupation of Japan, issued a thousand formal orders in the first ten months of the Occupation. These were sent to the imperial government, which then issued orders to the prefectures for implementation. Very little "American interference at local levels" occurred.[7] The medical studies of the survivors, on the other hand, required intense engagement between Americans and Japanese at several levels. Yet, like the orders of the Occupation, the medical studies depended on implementation by local Japanese.

I want to suggest that the American studies of the survivors were a form of science comparable to colonial indirect rule, in which existing systems of administration (in this case data collection systems) were maintained, but overlaid with a level of colonial control.[8] The term "colonial science" has been used to describe science as practiced in the colonies by Westerners (Kumar 1990), science as racist moral justification for imperialistic practices (Curtin 1990), science affected by colonial ideologies of control (Secord 1982), and science as practiced in the colonies by local scientists (Shiva and Bandyopadhay 1980). I here suggest a different meaning for the term, that is, colonial science as science, conducted by outsiders, that depends on local knowledge, particularly when that knowledge is invisible to the colonizers themselves.

7. Ralph Braibanti explores the relationships between American and Japanese institutions in "The MacArthur Shogunate in Allied Guise" (1984, 82).

8. I am indebted to Henrika Kuklick for her insights on this question. See particularly her recent work, *The Savage Within: The Social History of British Anthropology, 1885–1945* (1991, 280–86).

THE JOINT COMMISSION

The idea that a long-term study of the biological impact of the bombs' radiation might be appropriate occurred to more than one observer almost immediately after the bombings in early August 1945. By mid-September, less than six weeks after the bombings, three American teams and two Japanese teams had been appointed to study the biological impact of the radiation released by the atomic bombs.

Japanese physicians began to survey the medical effects of the bombings within days of the attacks. Military physicians from the Army Medical College and the Tokyo First Army Hospital arrived in Hiroshima on 8 August, two days after the bombing (Oughterson and Warren 1956, 439). After Japan's capitulation on 14 August, work accelerated in both Hiroshima and Nagasaki. Physicians in the two cities began keeping records on all the injured and their families. By 29 August investigative medical teams from Kyushu Imperial University arrived in Nagasaki and began interviewing survivors and documenting injuries. A similar group from Tokyo Imperial University went to Hiroshima at about the same time.[9]

Working conditions were extremely poor. A typhoon interrupted medical studies and relief measures in Hiroshima on 17 September, setting off a landslide that destroyed Ono Hospital, "carrying 10 investigators [from Kyoto University] and most of the patients, records and equipment into the sea" (Oughterson and Warren 1956, 439). Medical supplies were unavailable. One physician near Nagasaki recalled doing surgery without rubber gloves. He was embarrassed when some American doctors came by and were "very surprised." "They asked 'Do you always do surgery without gloves?' I had to explain that we did not have any gloves."[10] In addition, medical personnel were desperately in demand. Of necessity, emergency relief work took precedence over the collection of scientific data.

9. Japanese physicists also arrived in the city and quickly determined both the nature of the bomb and the height at which it detonated. By simple triangulation, based on the shadows burned into various walls and structures around the city, it was possible to estimate how high the explosion had occurred. American scientists considered this height a military secret for most of the next year, but it was widely known by Japanese scientists within a week of the bombings (Liebow 1965, 77).

10. Raisuko Shirabe, oral history interview, 22 October 1981, quoted in chapter 2 of an unpublished and unidentified manuscript fragment, GT, 9. I believe this manuscript is by John Bowers. It is dated 12 February 1987 and is presumably a portion of the manuscript prepared at NAS request by historian Bowers, as an institutional history of the ABCC. The draft was never approved for publication, and the NAS has refused to allow me or other researchers interested in the ABCC to have access to it.

On 14 September, the National Research Council of Japan set up a Special Committee for the Investigation of the Effects of the Atomic Bomb. The committee was to study not only medical effects but also building damage, fire damage, atmospheric effects, and effects on plants and wildlife. Its chair was Hayashi Haruo, a physician who taught at Tokyo Imperial University and also served as president of the National Research Council of Japan. The director of the medical section was Tsuzuki Masao, who was later to play an important role in Japanese relations with the ABCC.[11] Under this plan, about 150 research workers and a thousand assisting personnel came to Hiroshima and Nagasaki to begin the study and to help with rescue work (National Research Council of Japan 1953).

Meanwhile, by early September 1945, the American military created three medical teams to study the survivors: an Army team, headed by Ashley "Scotty" Oughterson; a Navy team, led by Shields Warren; and a Manhattan District (U.S. Army Corps of Engineers) team, directed by Stafford Warren (the two Warrens were not related). These three groups were not yet working together—indeed, they were "all over the place"—and there had been no attempt to unify them into, as they would later be called, a joint commission.[12] According to Shields Warren's later account, Navy surgeon general Ross McIntyre appointed the first team after biomedical staff in the Manhattan District complained that physicist Robert Oppenheimer was not interested in a medical study of the survivors.[13] Then a consultant to McIntyre, Shields Warren urged the Navy to constitute its own team. A medical study was "very urgent," he said, and the United States had a "grave responsibility" to get "competent medical teams into the area." McIntyre was sympathetic and told Warren to pick a team from the Naval Medical Research Institute at Bethesda, Maryland.[14] When Norman T. Kirk, surgeon general of the Army, learned that the Navy

11. Tsuzuki was fluent in English. He had studied radiation biology at the University of Pennsylvania in 1925 and 1926. He died of lung cancer in April 1961. There is an interview with Tsuzuki's son Masakazu in the manuscript cited above (Bowers?).

12. Shields Warren, oral history interview with Peter D. Olch, 10–11 October 1972, Boston, NLM.

13. Several frustrated biomedical members of the Manhattan District sought out Shields Warren shortly after the bombings to complain that the district planned no medical study. Warren blamed Oppenheimer, saying the physicist had "contempt for the inexact experiments of the life sciences." Shields Warren to Seymour Jablon, 10 May 1973, drawer 24, NAS. He also said Oppenheimer had a "deep antipathy for doctors," whom he regarded as "very unscientific." Peter D. Olch with Shields Warren, oral history interview, 10–11 October 1972, Boston, NLM.

14. McIntyre attached this team to the Naval Technical Mission to Japan—an affiliation which simplified dealings with the Army occupation—and sent them off to Okinawa to wait for armistice. Shields Warren, oral history interview, 10–11 October 1972, Boston, NLM.

had appointed a medical team, he appointed one as well.[15] Oughterson, who directed the Army team, was the Army surgical consultant in the Pacific theater and an associate professor of surgery on leave from the Yale Medical School.[16]

Around the same time, General Leslie Groves of the Manhattan District "became aware of the rumors" of the military teams and added a medical team to the group already planning to survey the physical consequences of the attack at Hiroshima under Brigadier General Thomas F. Farrell. Stafford Warren, who had played a role in Manhattan District radiation studies on mice, was appointed to direct this group.[17] Thus, apparently for competitive reasons, three separate American medical teams set out to assess the biological effects of radiation on the survivors.

Access to the two bombed cities in the early fall of 1945 was essentially uncontrolled, but transportation was a problem. Japanese trains were crowded, slow, and perceived as dangerous for Americans. The medical teams had to travel to Hiroshima and Nagasaki either by ship or by air, which meant they had to wait until such transportation was made available by the Occupation bureaucracy, the Supreme Commander of the Allied Powers (SCAP). As a consequence, Shields Warren and the Navy team waited for several weeks in Okinawa, not arriving in Nagasaki until 29 September. Army and Manhattan District teams were similarly held up in Tokyo until 12 October, though Oughterson and Stafford Warren themselves accompanied a special survey to Hiroshima around 5 September, under the direction of Farrell (Liebow 1965, 99).[18] On 12 October, MacArthur ordered all three teams to cooperate as a

15. Liebow's account of the appointment of the Army team suggests instead that Ashley Oughterson came up with the idea of an Army medical study of the survivors independently, while he was aboard a ship in the Pacific, but both these accounts were produced many years after the events in question (1965, 78).

16. Oughterson was on board the S.S. General Sturgis in the Pacific by 28 August, busily preparing to study the survivors. He "hoped that the Japanese may have already organized an investigation of the casualties" but felt it "unlikely under the circumstances." Oughterson's memo on board the Sturgis is reprinted in Averill Liebow's diary (Liebow 1965, 82–83).

17. Stafford Warren and Shields Warren were not related. Stafford Warren was chairman of the radiology department at the University of Rochester School of Medicine and Dentistry. During the war he was a consultant to the Manhattan Engineering District. He directed the University of Rochester radiation research project, begun in April 1943. Later the same year he was commissioned a colonel in the Army Medical Corps. *Biological Effects of External Radiation* includes introductory discussion of Stafford Warren's role in the University of Rochester project (Blair 1954).

18. A nonmedical group sent in to assess the risks of residual radiation was in Hiroshima in early September, under Brigadier General Thomas Farrell. Farrell's medical officer, Stafford War-

joint commission under the chairmanship of the Army group's leader, Ashley Oughterson. The SCAP order came through just in time to go aboard a flight bringing part of the group to Hiroshima (Liebow 1965, 135). The business of fieldwork led to impromptu cooperation with Japanese teams, and in the end the Joint Commission included Army, Navy, and Manhattan District physicians and two Japanese groups.[19]

One of the first orders of business for the American physicians was to find out what the Japanese were already doing. Accordingly, on 21 September Oughterson met in Tokyo with Tsuzuki, the physician in charge of Japanese efforts to study the biological effects of radiation on the survivors.

TSUZUKI MASAO

Tsuzuki was Japan's leading authority on the biological effects of radiation. He was an extremely powerful scientist, then fifty-three years old, whose cooperation was crucial to the American project. He had begun exploring the biological effects of radiation in the 1920s, working with rabbits exposed to X rays.[20] By 1939 he was rear admiral of the Naval Medical Corps, held a professorship at Tokyo Imperial University, where he performed lung surgery, and served as head of the medical section of the Japanese National Research Council.[21] When the bombings occurred, he immediately began efforts to assess their biological effects. In late August he visited Hiroshima and examined patients suffering from "A-bomb sickness." He concluded that there were four causes of injury in the bombed city: heat, blast, primary radiation, and "radioactive poisonous gas." Tsuzuki characterized the gas as containing strontium and other radioactive substances, which when inhaled would be deposited in the bone marrow. He planned to send out international reports on the existence of

ren, met with Ashley Oughterson on 4 September (in Tokyo) to discuss a coordinated effort. The Hiroshima team was held up in Tokyo by bad weather and transportation difficulties, including landslides that had damged the rail line. The Nagasaki group came by sea and was therefore able to reach Sasebo (Liebow 1965, 84, 128–31).

19. This collaborative organization set up headquarters in a rayon mill at Ujina, near Hiroshima, in mid-October (Liebow 1965, 72).

20. "Memory of Dr. Masao Tsuzuki," translation of article in *Chugoku Press,* 27 April 1961, drawer 2, Dr. Masao Tsuzuki, NAS.

21. Tsuzuki was born in Hyogo Prefecture in 1893. He graduated from the Tokyo Imperial University Medical School, then joined the navy, serving at naval hospitals in Ise and Kure. Tuzuki Masao, "Interview of Dr. Masao Tsuzuki by Dr. Ichiro Nakayama," translation of article in the *Yomiuri Press,* 16 August 1959, drawer 2, Dr. Masao Tsuzuki, NAS.

this poisonous gas, which he said violated the rules of the Hague Tribunal, but Occupation censorship apparently prevented this.[22]

Tsuzuki was an enigmatic character to many Americans. He was respected even by those who sometimes found him difficult or ambivalent (Schull 1990, 132), and he was too important in the Japanese scientific community to ignore. Some of his stature was captured in an effusive obituary, on his death in 1961, that compared him to Galileo, for standing up to Occupation authorities in his efforts to publish information about the medical effects of radiation, and to Giordano Bruno, for having been purged by SCAP.[23] As a former admiral in the Imperial Naval Medical Corps, Tsuzuki was indeed purged in 1947. American officials banned him from holding a position with the National Research Council or serving on any of its committees and eliminated his position at Tokyo Imperial University. But American scientists appealed to Norman T. Kirk, surgeon general of the U.S. Army, to allow Tsuzuki to continue his work on the biological effects of radiation under American supervision.[24]

Later Tsuzuki claimed in newspaper interviews that American scientists published Japanese work under their own names and that Japanese scientists working with American investigators had been treated unfairly. In one published account he said Japanese scientists had done more work on the biological effects of radiation than the scientists of any other nation, but since it was carried out during the Occupation "most of the work was published in the names of Americans and some in the name of Englishmen." In the same report he suggested that Japanese scientists working with the Joint Commission, and later with the ABCC, were not properly fed. The Americans were eating "plenty of beef steaks" while the Japanese staff lived on rationed food.[25]

Yet despite his public criticisms and open skepticism about American intentions, he was an important, politically sophisticated contact and a poten-

22. Presumably he was simply referring to radioactive particles that could be inhaled. While American military officials would readily concede that the bomb caused deaths from heat, blast, and radiation, they were not willing to recognize the presence of "radioactive poisonous gas." Later, the ABCC's first director, Lieutenant Colonel Carl Tessmer sought to meet with Tsuzuki to discuss this claim which Tsuzuki continued to publicize even in the 1950s, but Tsuzuki refused. Nobutaka Ushio to Hiroshi Maki, "Confirmation by Dr. Tsuzuki," account of an interview with Tsuzuki, 17 August 1954, drawer 2. Dr. Masao Tsuzuki, NAS.

23. This obituary included the unlikely suggestion that the pulmonary cancer that killed Tsuzuki was the consequence of his having briefly visited Hiroshima in the weeks after the bombing. "Memory of Dr. Masao Tsuzuki," official translation of article in *Chugoku Press,* 27 April 1961, drawer 2, Dr. Masao Tsuzuki, NAS.

24. Lewis Weed to Norman T. Kirk, 27 March 1947, drawer 24, NAS.

25. "Interview of Dr. Masao Tsuzuki by Dr. Ichiro Nakayama," translation of article in the *Yomiuri Press,* 16 August 1959, drawer 2, Dr. Masao Tsuzuki, NAS.

tially valuable ally. He was often helpful to the Americans and numerous times stated his basic support for the American studies, despite occasional disagreement with particular actions or policies.[26]

Tsuzuki's attitude in his first meeting with American scientists and physicians in October 1945 was conciliatory. He agreed to compile a list of all Japanese publications and studies underway and to assist Oughterson in preparing forms, in Japanese and English, to be used in tracking patients. He quickly assembled a group of twenty-five Japanese investigators to help the Americans in their work, and less than twenty-four hours after Oughterson's first appeal the Americans found themselves meeting with a Japanese research team in a conference room at Tokyo Imperial University.

This meeting was characterized by Japanese reserve and American enthusiasm. Oughterson gave a "little speech in which he emphasized that the war was over and that in any event science was nonpolitical." He said the Americans needed Japanese help, "not only because of the language barrier but because we needed the highly skilled medical scientists for which Japan was famous." He closed by saying that, while the Americans expected full cooperation from the Japanese, they would never "rob them of the fruits of their thought and work in publication."[27]

Tsuzuki's cooperation came despite the fact that he had already begun an ambitious study of the survivors himself. He organized a population survey in Hiroshima in August 1945, in an effort to create a data base for further studies. Volunteers walking through the city circulated questionnaires to 145,000 persons. The forms asked survivors to describe where they were at the time of the blast and to explain their injuries.[28]

By the time the Americans arrived, six to eight weeks after the bombings, many of the most seriously injured survivors had died. Since the American studies were intended to provide information for civil defense plans, the nature of these earlier deaths was of interest. American investigators needed to know how and why people died in the wake of an atomic attack. They wanted to

26. In a cordial letter to Robert Holmes in 1955, he stated that "in these ten years [since the bombings] I myself have been endeavoring to promote an intimate relationship between American and Japanese scientists, and to assist, both openly and secretly, the work of the ABCC as much as I could." Tsuzuki Masao to Robert Holmes, 10 January 1955, drawer 2, Dr. Masao Tsuzuki, NAS.

27. See the description of this meeting in Liebow's published diary (1965, 101–2). Within a few days the Americans had assigned fourteen of the young Japanese researchers who attended this meeting to Hiroshima and eleven to Nagasaki (Liebow 1965, 105). Nagasaki was believed to require fewer physicians because of the presence of a well-established research group at the naval hospital there.

28. See Atomic Bomb Casualty Commission, "Report No. 2, for the Period 2–7 December 1946," 6/A, JVN.

know how to triage radiation victims—"to be able to select those who have absorbed lethal amounts of irradiation so as to be able to concentrate on those whom one can hope to save"[29]—but to do so they needed information about the experiences of the most seriously injured survivors in the days and weeks immediately after the bombings. They were unsure what symptoms might most reliably be taken as signs of acute radiation sickness, and a central goal of these initial medical studies was to determine what severe radiation sickness looked like.[30]

Collecting information on those who had survived the bombing but died before the arrival of American teams was therefore a top priority. But it was "especially difficult" because the records were scattered around Japan. The health care delivery system was destroyed in both cities so, for example, many Nagasaki survivors were treated at Kyushu Imperial University hospital in Fukuoka. Others were hospitalized as far away as Osaka, Okayama, Kumamoto, and Tokyo. Survivors in Hiroshima had been treated in Kure, Kobe, Kyoto, Osaka, and Tokyo. The American medical teams had to check major university hospitals around Japan, and in this way they obtained the medical records of about six hundred survivors who had left the two cities seeking care but had died before the Americans arrived (Oughterson and Warren 1956, 437).

The military physicians tried to translate these records of the dead onto the standardized form that was to be used for analysis of survival rates. This form asked for information about the location, position, and shielding of the survivor at the moment of the bomb's detonation and for detailed descriptions of symptoms and their time of onset. The American teams plotted the locations of survivors using maps of Hiroshima and Nagasaki marked with seven equally-spaced rings centering on ground zero. Eight radial lines further divided each map; any location could therefore be specified according to distance (rings) and direction (radial lines) from the hypocenter. This scheme helped to illustrate both patterns of survival (how rates varied with distance from the hypocenter) and presumed levels of radiation exposure of survivors. Those already dead when the Americans arrived were placed within this

29. James V. Neel, "Memorandum for the Record," regarding a conference with Everett Evans, Herman Pearse, Shields Warren, Philip Owen, and James Neel, 26 August 1947, JVN.

30. They found that most persons with radiation sickness had experienced severe nausea and vomiting on the day of the bombing. These immediate symptoms were present even if other symptoms of radiation sickness—loss of hair, gingivitis, subcutaneous bleeding—did not appear for several weeks. Surprisingly, there seemed to be no clear relationship between burns and radiation sickness, but hair loss and subcutaneous bleeding were closely correlated with distance from the hypocenter. Investigators judged these the best two signs of high radiation exposure (Oughterson and Warren 1956, 166–73).

scheme, if possible, based on information provided by physicians or family members (Oughterson and Warren 1956, 462–64).

The detailed questions about location and position at the moment of the explosion were intended to provide a rough guide to radiation exposure levels, which were unknown. The estimated amount of radiation released by the bombs was a sensitive military secret in 1945.[31] Like most later analyses, the Joint Commission analysis estimated dosage in terms of distance from ground zero, rather than from the bomb itself. In both Hiroshima and Nagasaki, the bombs exploded several hundred meters above the ground. Individuals directly beneath the bomb, therefore, were not significantly closer to the source of radiation than those 300 meters away. The Joint Commission tried to take into account an individual's body position and shielding at the moment of detonation—even a pole or tree trunk could provide some protection from the bomb's radiation. The Americans did not include estimates of internal radiation, that is, inhaled or ingested radioactive particles, in their calculations. Nor did they include estimates of exposure to residual radiation, even for those near the hypocenter who might have remained in the area for some time after the bombings.[32] While some who came to the cities in the relief effort reported having suffered symptoms consistent with radiation sickness, the American military reports suggested that residual radiation levels were too low to have caused such sickness.[33]

When the American physicians began interviewing survivors, they faced

31. Radiation released could only be estimated by Manhattan Project physicists, who were in the process of determining these figures in the fall of 1945. In any case the information was considered a military secret—as was virtually everything about the bombings—and even as late as 1956 medical investigators did not have access to specific estimates. Instead, they used a report on radiation released by a "typical" atomic bomb, though of course different bombs were used in Hiroshima and Nagasaki, and neither was quite "typical." This report was Samuel Glasstone, *The Effects of Atomic Weapons,* produced by the Los Alamos Scientific Laboratory (1950)

32. For a detailed discussion of the dosimetry problem, see John A. Auxier, *Ichiban: Radiation Dosimetry for the Survivors of the Bombings of Hiroshima and Nagasaki* (1977).

33. On the Japanese response to residual radiation, see Shohno Naomi and Sakuma Kiyoshi's article, "The Fundamental Examination of the Amount of Radiation Received and Its Correlation to Acute Symptoms" (1958). Also, see Grant Taylor to Carl A. Harris, 7 August 1951, 6/S, JVN: "We have received several communications regarding the possibility of a residual radiation effect. One of these was sent to us through Dr. Davison and had to do with a young man in the United States Army who was detailed to help clean up Hiroshima. A similar case was that of a young United States soldier who operated a bulldozer and inhaled much dust during the cleaning up process. etc. etc. Both of these, as I understand it, now have leukemia. . . . Several months ago we began a study in an attempt to chase down the rumors of epilation, etc., which were said to have occurred following a visit to Hiroshima shortly after the explosion."

problems provoked by both the language barrier and the cultural distance between them and their subjects. None spoke fluent Japanese, so interpreters were necessary, but a more fundamental problem was that none were fluent in Japanese culture. Even the recording of so simple a detail as age could be complicated by differing cultural constructions. The Japanese traditionally considered a child to be "one year old" at birth, and all children born in a given year turned "two years old" on the following new year. Thus a child born a few hours before the new year would be two years old the next day, even though by American calculations the child was less than a day old. Since the Japanese calendar was unfamiliar to the Americans, it was easier to ask for age than for date of birth. The investigators therefore accepted this scheme, but felt it undermined their analysis of young children (Oughterson 1955, 3).

The most serious error in these interviews was a consequence of alphabetic confusion. Interviewers failed to record names of persons interviewed in the Chinese characters used for Japanese names, kanji. Rather, the Americans wrote down the names in Romaji, that is, transliterated into the Roman alphabet. The Joint Commission team members did not "sufficiently realize the difficulties of identifying persons by their recorded names in Japanese." Many kanji characters sound the same, though they look different. Such differences were impossible to represent in Romaji. In addition, three distinct systems of transliteration were in use during the early Occupation. The decision to record the names in Romaji may have reflected Occupation interest in language reform in Japan. Allied occupiers deemed the complicated writing system of kanji characters and two syllabaries, katakane and hiragana, antithetical to democracy and universal literacy. But since no formal Occupation statement on this reform effort appeared until February 1946, it is more likely that writing the names in Roman letters was simply easier for the Joint Commission investigators.[34]

The imprecision of these transliterations made it impossible for later investigators to compare radiation histories taken in the fall of 1945 by the Joint Commission with those taken years later by the ABCC, since the original interview subjects often could not be matched to survivors contacted by the ABCC. When Shields Warren himself returned for a brief survey in the summer of 1947, his group attempted to contact persons interviewed by the Joint Commission in 1945. Out of 300 persons examined, only 113 proved to be the same

---

34. The Occupation effort to eliminate kanji in favor of romanized Japanese writing, on the assumption that romanized writing was somehow relevant to democratization, is explored in Martha Ellen Hardesty's Ph.D. dissertation, "Language, Culture and Romaji Reform: A Communications Policy Failure of the Allied Occupation of Japan" (Hardesty 1986, 111–21, 128–54). See also Liebow (1965, 111) on the Joint Commission's use of Romaji.

individuals. Warren observed that "some Japanese names having the same sounds but expressed by different characters become confused when transliterated into Romaji, and when again translated into Japanese characters for the use of the police may result in the wrong persons being contacted."[35] A similar effort by Fred Snell, an ABCC physician in Hiroshima, found a match of only 5 percent.[36] To complicate things further, Joint Commission interviewers often did not record the home address of interview subjects. Accurate addresses might have made it easier to match up interview subjects. (Liebow 1965, 111).

Comparative radiation histories, drawing on both the Joint Commission interviews and the later ABCC interviews, could have provided insight into the relative consistency of survivors' stories over time. The ABCC staff was later frustrated by these early oversights, which complicated the problem of estimating dosimetry and assessing the efficiency and validity of the history-taking procedures.[37] The Joint Commission interviewers presumably did not realize that recording the names in Romaji would reduce the long-term scientific value of their interview data. Yet the case suggests how ignorance of (or indifference to) a cultural parameter—in this case the importance of using kanji—could affect the usefulness of the data.

The Joint Commission teams examined many patients still alive and hospitalized in October. But the Americans also wanted to examine and interview survivors who were not hospitalized. To facilitate this they opened clinics, in the Post Office Hospital in Hiroshima and in a hospital in Nagasaki. They invited "healthy survivors" to these clinics for free examinations by teams of Japanese doctors and nurses. The Joint Commission maintained these clinics for only about a month, until late November 1945 (Oughterson and Warren 1956, 440–41).

In a relatively brief period, from early October to mid-December 1945, the Joint Commission medical teams collected data on 6,993 survivors at Hiroshima and 6,898 at Nagasaki, most of whom had lived for at least twenty days after the blast. The results indicated that the blast, fires, and debris from the explosions killed more than 87 percent of all persons within one kilometer

35. Warren's report was "Preliminary Report of Investigation of Atomic Bomb Casualties in Hiroshima and Nagasaki, April, May and June 1947," Committee on Atomic Casualties, Organization, NAS.

36. In August 1947 Snell had the records of a Japanese investigator, Matsubayashi, transliterated into Romaji and then filed alphabetically, so that these records could be compared with the Joint Commission files. But without home addresses such matches were extremely difficult. See Fred Snell to James V. Neel, 11 August 1947, JVN.

37. "ABCC Progress Report," 15 April 1948, General Headquarters SCAP, Public Health and Welfare Section, JVN.

of the hypocenters immediately. Radiation produced the most serious injuries among those who survived in this very close group (Oughterson and Warren 1956, 29–71).

The medical teams were not the only American military investigators who made their way to Hiroshima and Nagasaki in the weeks after the bombings. Military officials were also interested in the physical effects of the bombs on buildings and other structures and materials. But the four teams sent to assess the physical effects of the bombings on buildings and other inanimate materials worked independently; made independent calculations of the bomb's height, location, and radiation mix; and did not pool their results (Hubbell, Jones, and Cheka 1969). These independent estimates later allowed for comparative studies of physical effects, which were not possible in the biological studies (Auxier 1977, 3–4). The biological and medical teams collaborated from the beginning, because their complicated work depended on contacting and examining a large number of patients, that is, on social interaction.[38]

The Americans took up a field of study already well populated with Japanese scientists, becoming collaborators and drawing on the local knowledge incorporated in the Japanese studies. When they returned to the United States, after less than four hectic months in the field, the medical questions raised by radiation exposure at Hiroshima and Nagasaki were again left to the Japanese investigators. For ten months, from January to November 1946, no American medical studies or surveys of any kind were underway at Hiroshima and Nagasaki.[39] Meanwhile, the Japanese continued their own work with the survivors.[40]

38. Gannon points out that teams of experts carried out their own separate investigations of various aspects of the bombs' effects—special kinds of physical damage, civilian defense, morale, community life, utilities and transportation, various industries, and general economic and political impact—in a period of ten weeks from October to December 1945 (United States Strategic Bombing Survey 1973, 1).

39. The Army issued a classified six-volume report on the Joint Commission findings in 1946 and published a summary in 1956 (Oughterson and Warren 1956).

40. In addition, Japanese geneticists I. Abe and M. Chine of the Japanese Special Committee for the Investigation of Atomic Bomb Effects collected *Drosophila* shortly after the bombings in both Hiroshima and Nagasaki, expecting to be able to use them for genetic studies. Unfortunately the work was discontinued because of lack of suitable fly media. Also, in Nagasaki, a breeder had independently collected rabbits exposed at 1.6 kilometers from the hypocenter, in the hope that they might be useful in genetic studies. These rabbits were mated but had to be destroyed for food before the young were born. Some studies of plants were also underway. On 8 May 1947 Dr. Takeo Furuno reported that many "twin onions" were being found in the immediate vicinity of the hypocenter. Potatoes, cabbages, onions, broad beans, mustard, lettuce, kale, and squash growing near the hypocenter were also examined. The twin-onion story appeared in *Stars and Stripes*, 15 May 1947. "These and other abnormalities of vegetables were ascribed to atomic bomb effects by the Nagasaki prefecture Agricultural Experts Association, according to the newspaper." All reports

They supplied American researchers with information and specimens (reluctantly), but they did not abandon their own work. Nor did they integrate their data with that of the Joint Commission. Such parallel efforts persisted in U.S.-Japanese collaboration throughout the Occupation (National Research Council of Japan 1953, iii).

While preparing their Joint Commission report, Oughterson, Shields Warren, and Stafford Warren decided a longer, more comprehensive American study was desirable. Their 15 May 1946 report recommended that a "permanent American Control Commission" supervise studies at Hiroshima and Nagasaki of possible hematological effects, genetic and fertility effects, and carcinogenic effects. They also noted the importance of a permanent registry of survivors.[41] The probable widespread use of atomic energy would make knowledge about the biological effects of radiation extremely important, they said, and the elite National Academy of Sciences (not the military) should organize a large-scale follow-up study of the survivors.[42]

THE ATOMIC BOMB CASUALTY COMMISSION

In late June 1946, Lewis Weed, chair of the NAS research arm, the National Research Council, called together a group of scientists to discuss the potential for such a study. The group agreed that a "detailed and long-range study of the biological and medical effects upon the human being" was "of the utmost importance to the United States and mankind in general." It recommended either that the NRC undertake the study or that the president create a federal "Commission on Atomic Bomb Casualties."[43] But the committee members recognized that a federal commission was unlikely; the alternative proposal,

---

of lost biological materials (including the rabbits) are in Atomic Bomb Casualty Commission, "Interim Report No. 7, Covering the Period 30 April to 7 May 1947," 6/A, JVN.

41. "Activities of the Atomic Bomb Casualty Commission," 12 June 1948, box 9, JCB.

42. On 15 May 1946 Colonel Ashley W. Oughterson addressed a letter to the surgeon general of the U.S. Army urging that the National Research Council be appointed to make recommendations regarding plans for a continuing study of the medical effects of the atomic bombs. Oughterson's suggestion was communicated to Lewis H. Weed, chair of the division of medical sciences of the NRC, on 28 May, and on 28 June Weed replied to the request. He had consulted by then with representatives of the Army, Navy, Public Health Service, State Department, American Cancer Society, and other groups. Lewis Weed to Norman T. Kirk, 28 June 1946, drawer 29, NAS.

43. This commission would consist of one representative each of the three surgeons general, the medical director of the Veterans Administration, and five or more civilian scientists in appropriate specialties.

for an NRC study, "even though less desirable as a long-term mechanism," seemed more feasible.[44]

Later some in the NAS (and in the Atomic Energy Commission) questioned the suitability of the NRC as sponsor of the study. The National Academy of Sciences, created in 1863 by a congressional charter approved by Abraham Lincoln, had long advised the federal government on scientific and technological questions, but it was not a government agency and did not receive government funds directly. Government officials generally requested that the Academy carry out a particular study, which was then funded through the appropriate federal agency. The National Research Council, created in 1916 is the research arm of the NAS, but traditionally NRC research has consisted of compiling and analyzing the scientific work of others to produce reports for use in setting policy. While this has been changing in recent years—with increasing NAS sponsorship of original research in transportation and medical care systems—in 1946 the proposal that the NAS carry out epidemiological research in radiation biology was unusual.[45] Despite the scientific credentials of its members, the NAS-NRC was ill equipped to take on such a study in a country halfway around the world, as later events made clear. Some within the NAS had doubts about the advisability of an American study in Occupied Japan from the beginning.[46] But the NAS's high scientific status made it an appropriate sponsor from a political perspective and made possible the involvement and endorsement of respected American scientists.[47]

44. Lewis Weed to Norman T. Kirk, 28 June 1946, drawer 29, NAS.

45. For a detailed description of the organization and history of the National Academy of Sciences, see a guidebook published by the academy, *Questions and Answers about the National Academy of Sciences, National Academy of Engineering, Institute of Medicine and National Research Council* (1989).

46. The most telling criticisms of the epidemiological methods used in the early somatic studies were contained in the preliminary and final reports of the group under the direction of University of Michigan epidemiologist Thomas Francis in 1955. See Ad Hoc Committee for Appraisal of ABCC Program to R. Keith Cannan, 6 November 1955, NAS. Detlev Bronk commented in a 20 October 1949 meeting of the Advisory Committee for Biology and Medicine of the U.S. Atomic Energy Commission that "the U.S. government operating in a foreign country is something I did not like from the first." See ACBM, "Minutes, 18th Meeting," 20 October 1949, RG 326, box 3218, ACBM, DOE.

47. Later proposals for control of the ABCC included turning the study over to the United Nations or assigning it to a major research university. For one detailed proposal to disentangle the NAS from involvement with the ABCC, see Robert Holmes to Keith Cannan, 10 February 1955, JVN. Holmes suggested that the project in Japan be turned over to the United Nations for staffing, even if the United States continued to supply all funding. "Solitary United States control of the project now is not as important as accomplishing the task at hand." Also, Shields Warren (representing a committee composed of Warren, John C. Bugher, Brigadier General Elbert DeCoursey,

The Atomic Bomb Casualty Commission was ordered into existence by Harry Truman in November 1946. Truman's approval, across the bottom of a letter from Secretary of the Navy James Forrestal, was the culmination of many months of work in both Japan and Washington.[48] This letter was actually written by Weed and Shields Warren. Weed's 28 June 1946 letter became the basis for the directive,[49] after Warren revised the text, removing Weed's tentative language and reducing the options to one: a National Research Council study.

In a remarkable omission, the letter listed the earlier researchers as consisting of Army, Navy, and Manhattan District teams, with no mention of the two Japanese teams and dozens of Japanese investigators. This may have reflected either Warren's low assessment of their importance to the project, or his perception that it would not be politically useful to emphasize Japanese contributions to the study in this context.[50] "Preliminary surveys involve about 14,000 Japanese who were exposed to the radiation of atomic fission," Warren noted. Such survivors offered a "unique opportunity for the study of the medical and biological effects of radiation which is of utmost importance to the United States." The study should "continue for a span of time as yet undeterminable" and was "beyond the scope of military and naval affairs, involving as it does humanity in general, not only in war but in anticipated problems of peaceful industry and agriculture."[51] Truman wrote "Approved" across the bottom of this letter, and signed and dated it, 26 November 1946.[52] This was the presidential mandate that created the ABCC.

A scientific study group for the proposed new organization met in Wash-

---

and Thomas Parran) recommended that the NAS-NRC turn over sponsorship of the ABCC to "another operational contractor, preferably of university type." Shields Warren to Detlev Bronk and R. Keith Cannan, 18 April 1956, box 9, ABCC April–July 1956, JCB.

48. James Forrestal to Harry Truman, 18 November 1946, box 9, ABCC January–March 1956, JCB.

49. Lewis Weed to Norman T. Kirk, 28 June 1946, drawer 29, NAS.

50. In other settings, Warren had suggested that the Japanese could not make significant contributions except under American guidance. See Shields Warren, "Preliminary Report of Investigation of Atomic Bomb Casualties in Hiroshima and Nagasaki, April, May and June 1947," 28 May 1946, Committee on Atomic Casualties, Organization, NAS.

51. James Forrestal to Harry Truman, 18 November 1946, box 9, ABCC January–March 1956, JCB.

52. See Shields Warren to Seymour Jablon, 10 May 1973, drawer 24, NAS (note: Warren misdates the letter as November 1947 instead of 1946); also Paul Henshaw to James V. Neel, 7 February 1947, JVN; and Lewis Weed to Norman T. Kirk, 28 June 1946, drawer 29, NAS. Neel later carried around a photocopy of this signed letter, to be produced whenever he needed more cooperation from Occupation officials. James V. Neel, interview by author, November 1988.

ington on 4 November 1946, just three weeks before Truman's official approval. The key members of this group were Weed, NRC physicians Austin M. Brues and Paul Henshaw,[53] and two Army representatives appointed to help organize and manage the study, First Lieutenant Melvin A. Block and Lieutenant James V. Neel, a medical doctor with a Ph.D. in genetics and one of the most important organizers of the ABCC.[54]

This group was thinking in grand terms, of a study that would last "several decades or longer" and would include experimental work with wild mice or fruit flies collected from the bombed cities. Expectations of cooperation from survivors were high at this early meeting, though the committee recognized that "recovered persons will probably prefer to forget their tragic experience rather than submit themselves for examination."[55]

Financial support for the study was to come from the new Atomic Energy Commission (the metamorphosed Manhattan Project) but not until late 1947. Brues and Henshaw interpreted the presidential directive as permitting support from the Army, Navy, Atomic Energy Commission, and perhaps other federal agencies. Yet military support was ruled out by Bureau of the Budget stipulations precluding external grants. The AEC, already heavily involved in supporting radiation research, was therefore the only potential government sponsor.

NRC chairman Weed met with Carroll Wilson, business manager for the AEC, in early 1947 to discuss possible support. The AEC had provided $75,000 for the publication of the Joint Commission's results and had a keen interest in the survivors. Wilson "expressed a desire to provide all possible means of support" to the new ABCC.[56] But the AEC's formal approval of fund-

53. Henshaw was a biophysicist with the Atomic Energy Commission who had been principal biologist for the Manhattan Project in Chicago during the war. Brues was a Harvard Medical School graduate (1930) who was then director of the biological division at Argonne National Laboratory. He had been with the Office of Scientific Research and Development in Boston from 1941 to 1944, and was chairman of the subcommittee on radiobiology of the NAS. He later worked with the National Council on Radiation Protection, the private council whose reports were published by the National Bureau of Standards and whose judgments set standards for occupational exposure to radiation. See biographical data in *American Men and Women of Science* (1962).

54. Also attending were James P. Cooney, commanding officer of the medical section, Manhattan Engineer District; George Dowling, chief of the research division of the U.S. Navy Bureau of Research and Development; William S. Stone, chairman of the Army Medical Research and Development Board of the Office of the Surgeon General; and two representatives of the War Department.

55. Paul S. Henshaw, "Report," 14 November 1946, JVN.

56. Committee on Atomic Casualties, "Minutes," 1 May 1947, NAS.

ing did not come until a good deal of groundwork for the study had already been carried out at the expense of the National Academy of Sciences and the U.S. Army.

CONCLUSION

The medical studies of the survivors—first by the Joint Commission, subsequently by the ABCC—constituted a hybrid culture, a local point at which American-Japanese interaction permitted knowledge and influence to flow both ways. The Americans incorporated many aspects of Japanese scientific studies into their own studies. They borrowed the cultural knowledge built in to Japanese surveys, adapting Japanese data collection systems to their own needs. They needed Japanese input in planning the wording of questionnaires, handling patient examinations, and dealing with the Japanese community. They needed the cooperation of a large number of Japanese—physicians, technicians and support staff, midwives, and survivors—as employees and study participants.

I would suggest that the Japanese input was substantial and far more important than many of the Americans—and even many of the Japanese—recognized. Japanese critics of the ABCC consistently suggested that the ABCC was a joint undertaking on paper only; the Japanese role was often unnoticed, invisible, and, from the perspective of both American promoters of the ABCC and Japanese critics, not very important. But the Japanese who worked with the Joint Commission and for the ABCC contributed cultural knowledge, social skills, data collection systems, and field practices that were critical to the scientific analysis. That neither side recognized this contribution suggests a shared interpretation of science as an abstracted data set, a unique cultural product with no culture. "The program of the ABCC is, on paper at least, a joint Japanese-American undertaking," noted the NRC's Philip Owen in a 1950 memo assessing the organization's "implications in international relations."[57] He expressed the common American belief that the program would be "guided and controlled" by Americans but promoted as a joint venture for public relations purposes. But if science is a field practice and a social construct, then the Japanese workers who collected data were active players. They shaped the

---

57. Philip Owen to Robert Keith Cannan, "Memo," 7 July 1950, Unitarian Medical Education Mission to Japan—1951, NAS. Also see Shields Warren, "Preliminary Report of Investigation of Atomic Bomb Casualties in Hiroshima and Nagasaki, April, May and June, 1947," Committee on Atomic Casualties, Organization, NAS.

research at its most fundamental level, in the day-to-day business of working in the field.[58]

On 22 November 1946, four days before Truman actually approved the program, and before any funding for a long-term study had been arranged, the members of the new "Atomic Bomb Casualty Commission" began a three-day journey, by air, to Tokyo. Neel, Henshaw, Brues, and Block then embarked on a six-week program of touring, socializing, and planning.[59] Two SCAP officials accompanied them to Kyoto, Osaka, Kure, Hiroshima, Fukuoka, and Nagasaki. One purpose of this tour was to select a control city—preferably a city that had suffered conventional bombing in the war, was roughly equivalent in size to Hiroshima, and was geographically convenient.[60] They maintained a dizzying schedule of luncheons, meetings, tours, and dinners. They did some sightseeing and gained a shocking first-hand understanding of the difficulties of carrying out a scientific study in postwar Japan. In this first field trip the ABCC team began the complicated process of constructing the ABCC as apolitical diplomacy—the diplomacy of scientific internationalism.[61]

58. For related work on colonial science, see Nathan Reingold and Marc Rothenberg, *Scientific Colonialism: A Cross-cultural Comparison* (1987).

59. A fifth member of the team, Navy Lieutenant Frederick Ullrich, was hospitalized off and on during the visit to Japan and returned early to the United States, having played a minimal role in the initial tour. See Atomic Bomb Casualty Commission, interim reports, November 1946–April 1947, JVN.

60. Ibid.

61. As John Beatty has pointed out, the apparently universal nature of science—its distance from politics—can make it a political asset. Science is "above" politics and therefore a useful diplomatic tool. See Beatty, "Scientific Collaboration, Internationalism, and Diplomacy: The Case of the Atomic Bomb Casualty Commission" (1993, 205–31, esp. 206–9, 215–19).

CHAPTER THREE

Into the Field

When they arrived in Occupied Japan, Americans found themselves in a complex culture they had difficulty interpreting. The Japanese were the subject of racist propaganda during the war—the U.S. War Relocation Authority had imprisoned Japanese-Americans in concentration camps during World War II (Drinnon 1987)—and many Americans had an image of the Japanese as subhuman.[1] At the same time, the Japanese had absorbed their

---

1. John Dower has explored how the racist images of the war years quickly shifted in the postwar Occupation period, the vicious Japanese ape (the "missing link" or King Kong monster) of the war years transformed, for example, to an organ-grinder's monkey (the charming mimic) in editorial cartoons and journalists' accounts. The Japanese, who had been mindless followers of a fascist state during the war, thus became docile pupils in the imagery of the Occupation (Dower 1986, 302–7).

own propaganda. Nagasaki physicians awaiting the first Americans in September 1945 urged their nurses to hide in the hills. They feared the Americans would rape all the women in the city (Akizuki 1981).[2] Both sides therefore faced the formidable problem of communicating across a cultural gap with a treacherous undercurrent of racism.

The members of the new ABCC team hoped that science could bridge this cultural gap. In their accounts of their early interactions with the Japanese community, they commonly portrayed their scientific study as a neutral ground, an open space within which the divisions of war, culture, and gender dissolved. They reported that meetings with Japanese physicians began tensely but ended in a spirit of warm cooperation, that Japanese midwives were at first indifferent or hostile but then saw the value of the American work and pledged to help, that leading Japanese scientists were at first suspicious of American motives but later became enthusiastic. Through such accounts, they constructed the ABCC as the equivalent of a demilitarized zone—a forum in which the war did not exist or did not matter. In this initial foray into the field, one of their guiding principles was that science could be a language above language, untethered to the details of history and culture.

Yet science in practice was not a universal language. As they quickly discovered, Japanese science was different from American science, a fact that made the visiting Americans uncomfortable, partly, I suspect, because it raised an implicit question: If Japanese science was culture-bound, was American science as well? British biologist J. B. S. Haldane once compared the ABCC to a "joint Roman-Jewish study of the physiological effects of crucifixion" at the time of Christ's death. Many biologists, he said, would have found such collaboration problematic (Haldane 1964). His comment captured an underlying tension in the premise of the ABCC, that is, that science could make irrelevant the circumstances that caused the biological events under study.

2. Akizuki was astonished when the first American physician, an ophthalmologist examining eye injuries in survivors, seemed to be kind and considerate. Another Nagasaki physician, Raisuke Shirabe, urged his nurses to stay but could not convince them that they would be safe. "On the night of August 17 [1945] I was caring for an old injured man when a Mr. Kataoka, the head of the machi [city block] visited and told me to stop medical therapy because there was a rumor that the American soldiers were coming. All the young girls were said to be fleeing to the mountains for protection. He also urged me to evacuate, but I ignored this suggestion and continued to administer therapy. So on the 18th most of the residents in the neighborhood had disappeared and only my family remained in Nameshi. . . . Eight nurses working for me were frightened and asked me to let them leave. I tried to convince them that there would be no trouble if the Americans came, but they insisted and left. I decided to close my first-aid station since there was no one to care for." Raisuko Shirabe, oral history interview, 22 October 1981 (Bowers?), cited in unpublished manuscript fragment, GT.

The ABCC team found that it was in fact extremely difficult to disconnect the medical effects at Hiroshima and Nagasaki from the war, the bomb, and the reality of American military occupation. SCAP's existence placed them in the uncomfortable role of representing a foreign military government; SCAP's goal of democratization dictated their relationship to the Japanese scientific community; and SCAP's censorship of Japanese scientific publication conflicted with their stated position that the Japanese would be equal collaborators in the study of the survivors.

VICTORS AND VICTIMS

On their second day in Tokyo, the ABCC team made the necessary visit to Tsuzuki Masao at Tokyo Imperial University. Tsuzuki set the tone that would characterize their later relations: His office at 9 A.M. was uncomfortably cold ("inside temperature 50 degrees F. or below") and his message was equally chilly. He outlined work then underway on the biological effects of radiation, pointing out that the Japanese research program had already produced 104 scientific reports on the topic.[3] He proudly presented the Americans with copies of some of these (unpublished) reports, on fertility effects and developmental abnormalities in children exposed to radiation in utero. He stressed the effectiveness of the existing Japanese study and noted that he had a thirty-thousand-yen appropriation from the Japanese National Research Council (approximately two thousand dollars, "but reported to be worth one-fourth this amount"). The ABCC team members, in their report on this meeting, dryly noted an "eagerness on the part of the Japanese to work on atomic bomb problems"—not necessarily with American help.[4]

It was only one of dozens of stops—diplomatic, scientific, and administrative—for the ABCC team in its first foray into the field in the winter of 1946. In the course of their initial work in Japan, Block, Brues, Henshaw, and Neel met with many Japanese officials, scientists, physicians, and others, including

3. Monica Braw mentions 80 Japanese reports by the end of 1946, while the ABCC report mentions 104. A total of 133 Japanese reports were listed in the records of the Joint Commission. See Braw, *The Atomic Bomb Suppressed: American Censorship in Japan 1945–1949* (1986, 120–29); also, Atomic Bomb Casualty Commission, "Report No. 1, for the Period November 22–30, 1946," 6/A, JVN; and Oughterson and Warren (1956, 444–46). Japanese scientists interested in publishing reports about atomic bomb effects complained to the first ABCC team in December 1946 about SCAP censorship policies. "Summary of General Impression," filed with Atomic Bomb Casualty Commission, "Report No. 3, for the Period 8–14 December 1946," 6/A, JVN.

4. Atomic Bomb Casualty Commission, "Report No. 1, for the Period November 22–30, 1946," 6/A, JVN.

the Japanese commanding officer for medical supplies, the president of the Japanese National Research Council, the director of the medical bureau of the Public Welfare Ministry of Japan, and several hospital staff groups.

Their accounts of these meetings often focused on the enthusiasm of their Japanese contacts for an American-sponsored study of the survivors. Thus Brues and Henshaw reported that a meeting with Japanese physicians in Nagasaki was "strained at first"—many of the hospital staff bore the scars of their own exposure to the bomb—but that tensions eased when the ABCC group explained their interest in Japanese involvement in the study. "The sincerity of our group was felt," they reported, and the Japanese physicians said this was "the first time since the war that Japanese and Americans had sat down together as scientists." These Japanese hospital staff members told the committee members "personal experiences" of the bombing and offered themselves as suitable subjects for study. The meeting ended on "what appeared to be a note of complete accord and anticipation of a fruitful cooperative study."[5]

Similarly, their meeting with a group of obstetricians, gynecologists, and midwives in Hiroshima began badly but quickly turned around. "At first it seemed we were not making headway," Brues and Henshaw reported. They discussed the ABCC's need for a complete record of all births, normal and abnormal, and gradually found that their audience was warming up. "Soon these people were relating their experiences and making plans. When the meeting was over, the ladies stood in line and bowed as we drove away in our jeeps."[6]

The Americans were interested in learning about all ongoing Japanese scientific work that might be relevant to the ABCC's program.[7] In Kyoto, they

5. Somewhat later, in March of 1947, James Neel spoke to a graduating class of nurses at the Hiroshima Red Cross Hospital. This was "the last class of nurses who were students at the time of the bombing. There were 200 in the entering class—70 graduated. It was a charged moment." James V. Neel to Austin Brues and Paul Henshaw, 31 March 1947, 6/J, JVN. Also see Atomic Bomb Casualty Commission, "Report No. 3, for the Period 8–14 December 1946," 6/A, JVN.

6. In a recent published memoir, Henshaw described this meeting in an almost identical narrative: "We met a group of sedate Japanese ladies in black kimonos, sitting in straight chairs with their hands folded neatly in their laps. Not a word was being spoken." A comment by a Japanese physician present, however, "broke the ice" and "we were on our way. . . . After a nice visit, the ladies stood in a line outside the cottage waving as we drove away in our jeeps." See Henshaw, "The ABCs of the Early Days" *RERF Update,* winter 1991–92, 12–13. The archival record is Atomic Bomb Casualty Commission, "Report No. 4, for the Period 16–22 December 1946," 6/A, JVN.

7. In Osaka, they met physics professor Asada Tsunesaburo, who had helped the Japanese determine the nature of the bomb, its altitude at detonation, the location of ground zero, and the

examined photographs of bomb victims and met with pathologists at the Kyoto Hospital and others elsewhere who were studying keloids, the large growths of scar tissue common among survivors. They found that shortages of ordinary supplies were hampering research. The Japanese had collected fruit flies—*Drosophilia melanogaster,* the great research organism of twentieth-century genetics—from the two bombed areas, to be used in assessing the genetic effects of the bomb's radiation. Some crosses were completed but the work was discontinued for want of necessary supplies: mashed bananas to feed the flies and known test stocks to serve as controls.[8] An attempt to study rabbits exposed at 1.6 km from the hypocenter ended for similar reasons; the rabbits were mated but had to be destroyed for food before the young were born. And they heard rumors of genetic damage—of orphans with light hair and blue eyes in Nagasaki, of plants with albino areas on leaves, and so on.[9]

On 6 December 1946, almost five years to the day after the Japanese attack on Pearl Harbor brought the United States into the war, the ABCC team went to Hiroshima and dined near the ruins of the city hall. The assistant mayor, Yamamoto Hisao, and a local physician, Matsubayashi Ikuzo, were their hosts. Together the Americans and the Japanese sat down to a three-course meal in a newly constructed building not far from the hypocenter. The assistant mayor apologized for the humble circumstances; the Americans were surprised when three main courses were served for lunch.[10] They learned that Matsubayashi and Kano Masato, a statistician, had begun a survey of survivors on the day of the bombing, largely at Tsuzuki's suggestion. Its purpose was to obtain a record of all surviving persons, where they were at the time of the blast, and the gen-

---

critical mass of uranium 235 in the days immediately after the bombing. The figures he provided in this meeting were placed in the confidential files of the commission. Asada said Japanese naval officials, hearing his initial reports of the cause of the destruction at Hiroshima, urged him to build a Japanese atomic bomb. They ordered him to select a secluded underground site and a group of scientists, and to produce an atomic bomb in six months. He responded that this would be folly, and, according to Asada, his arguments led the officials to concede that it was time for Japan to accept the Potsdam Declaration. Atomic Bomb Casualty Commission, "Report No. 2, for the Period 2–7 December 1946, " 6/A, JVN.

8. Atomic Bomb Casualty Commission, "Report No. 2, for the Period 2–7 December 1946," 6/A, JVN.

9. On light hair and blue eyes in Nagasaki, see Atomic Bomb Casualty Commission, "Report No. 3, for the Period 8–14 December 1946," 6/A, JVN; for plants with albino areas on the leaves, Atomic Bomb Casualty Commission, "Report No. 4, for the Period 16–22 December 1946," 6/A, JVN.

10. Atomic Bomb Casualty Commission, "Report No. 2, for the Period 2–7 December 1946," 6/A, JVN.

eral character of their injuries, if any. After more than a year of work, Matsu-bayashi had records on 145,000 of the 200,000 persons living in Hiroshima, based on questionnaires circulated by volunteer workers in each block of the city. Because of a lack of filing facilities, the records were arranged on tables around a large room, but they were neatly separated in piles corresponding to distance from ground zero.

Brues and Henshaw detected no reluctance on the part of these Japanese researchers to share their data. They were "unusually cordial and generous" and eager to collaborate, the Americans said. "They displayed what appeared to be all of their research material and gave us English translations of all their studies on atomic bomb injuries."[11] Later, Henshaw recalled that the "Japanese were eager to cooperate in the studies and they welcomed the assistance of American scientists."[12]

The ABCC team noted that the Japanese scientific observations "appear as accurate and reliable as could be expected in the light of circumstances, particularly the limited facilities." They particularly praised the work at Hiroshima, where volunteers had registered all survivors. "Though the accuracy of such records can be questioned, they are obviously of great value and serve admirably to keep up public interest in atomic bomb effects. They serve also, however to cause people to attribute certain conditions incorrectly to the atomic bomb."[13] This comment reflected a growing suspicion that the Japanese might be unable to contribute equally to the studies of the survivors.

BAD SCIENCE

Shields Warren observed in 1947 that while the Japanese investigators were "cooperative, interested and anxious to aid in the work of the Commission," their help would not be as "effective as might be desired owing to certain difficulties." These difficulties, Warren said, included the language barrier, inadequate medical training of many Japanese physicians, "lack of familiarity with the experimental method," and a "failure to appreciate the necessity for adequate controls." Warren said the Japanese were "alert and interested workers"

11. Atomic Bomb Casualty Commission, "Report No. 2, for the Period 2–7 December 1946," 6/A, JVN.

12. 25 March 1947, at the first meeting of the Committee on Atomic Casualties. Committee on Atomic Casualties, "Minutes," 25 March 1947, NAS.

13. See "Summary of General Impressions," filed with Atomic Bomb Casualty Commission, "Report No. 3, for the Period 8–14 December 1946," 6/A, JVN.

but that they worked "most effectively when associated directly with American investigators." [14] John C. Bugher, later director of the AEC Division of Biology and Medicine, also dismissed the Japanese contribution and suggested that the Japanese scientists could not be trusted. [15]

Many on the ABCC staff were also skeptical of Japanese scientific ability; both Carl Tessmer and Grant Taylor told a visiting consultant in 1950 that "the Japanese do not record scientific data accurately." [16] An American physician working in Nagasaki in 1954, expressed his concerns about the biases of the Japanese scientific community: "Just the thought of what the Japanese would do if they had free unrestrained use of our data and what they might publish under the imprimatur of the ABCC gives me nightmares." [17] Similarly, ABCC staff member Ray Anderson complained in 1948 that because he was "not perfectly satisfied with the caliber of the Japanese doctors who will be working for us . . . I feel it will be absolutely necessary to have one or more bilingual doctors (American or Japanese—well-trained by American standards) to supervise the project in order to be certain that the data collected will be reliable." [18] Two NRC consultants visiting the ABCC in 1948 suggested that a "good plastic surgery team" spend six weeks in Hiroshima and Nagasaki doing surgery on some of the bomb survivors. This would "teach the Japanese how to do good plastic surgery." The consultants were impressed with the "extreme deficiency in three fields, namely surgery, anesthesiology and radiology. The

14. Shields Warren, "Preliminary Report of Investigation of Atomic Bomb Casualties in Hiroshima and Nagasaki, April, May and June 1947," Committee on Atomic Casualties, Organization, NAS.

15. Or, as Shields Warren noted in 1947, that they "would require close supervision." Committee on Atomic Casualties, "Second Meeting," *Bulletin of Atomic Casualties,* 1 May 1947, NAS, 14. See also John C. Bugher, "Supplement to Report of the Atomic Bomb Casualty Commission Field Operations, 27 March to 7 May 1951" and "Report," drawer 28, Dr. John C. Bugher (AEC) Report on ABCC Visit—March–May 1951, NAS.

16. George A. Hardie, "Report of a Visit to ABCC in Japan, June–July 1950 [Copy No. 3]," NAS, 18.

17. James K. Scott was responding to a rumored plan to turn over the Nagasaki ABCC entirely to Japanese investigators. James K. Scott to Charles L. Dunham, 14 October 1954, drawer 4, Atomic Energy Commission, 1951–1961, NAS.

18. See Ray Anderson to James V. Neel, 16 July 1948, 6/E, JVN. Anderson also felt that his lack of fluency in Japanese language and culture handicapped his ability to supervise Japanese staff members at the ABCC. "I feel that I can operate only on 50 percent efficiency (if that good). I feel that there is a big deficiency in not being able to chat with the Japanese doctors about medical subjects, and worst of all I cannot properly evaluate their abilities and reliability." Ray Anderson to James V. Neel, 7 and 12 September 1948, 6/E, JVN.

modern physiological approach to surgery problems is almost totally lacking in most centers. Anesthesia methods in vogue simulate those of America of about 50 years ago."[19]

Some Japanese researchers, for their part, depicted the American study of the survivors as an extension of the military attack. Many physicians in Hiroshima and Nagasaki were themselves victims of the bombings, since in each city the hypocenter was near a large medical installation. Other Japanese scientists and physicians expressed a sense of being victimized a second time by American confiscation of blood and tissue samples. Shigeyasu Amano complained (much later, in 1959) that the ABCC performed autopsies of Japanese persons, but then "sen[t] all of its autopsy material back to the U.S. with the exception of a few specimens. It is absolutely impossible for Japanese scientists to make use of these specimens in their research."[20]

At first, the Americans working at the ABCC thought that these differences could be mitigated by the employment of nisei, that is, first-generation Americans born to Japanese parents in the United States. The ABCC favored nisei physicians and scientists in the hopes that they could help close the cultural gap between the Americans and the Japanese.[21] But several Japanese scientists announced in 1948 that they "did not want to work with Nisei and would not." And the director of the Japanese National Institute of Health, Kobayashi R., complained that the nisei "do not know the Japanese language well and they are not culturally adapted to the point where they can teach the Japanese scientists." He said the nisei had "haughty attitudes" and he "hoped [the ABCC] would not employ any more Nisei in the future."[22] Partly as a consequence of these complaints, by 1948 ABCC administrators decided to cease favoring nisei for positions. While American nisei continued to be hired by the

19. Everett Evans and Eugene P. Pendergrass, "Report to Dr. Philip Owen," 30 December 1948, drawer 28, Drs. Evans and Pendergrass Report on Visit to ABCC—1948, 10, NAS.

20. "When I was performing autopsies at Kyoto University, I gave some of my material to American research workers, but I have never been shown any materials collected by the ABCC. Despite the fact that ABCC is a research organization located in Japan, and is using Japanese people as autopsy material, it does not allow us to see a single histological specimen." Quoted in a translation of Imabori Seiji's *The Era of A- and H-bombs* (1959). The translation is filed in drawer 2, ABCC Press mid-1957–1964, NAS.

21. Grant Taylor persisted in this belief somewhat longer than other persons connected with the ABCC. "Dr. Taylor said the best thing that we could do with reference to personnel is to use bilingual Nisei." George A. Hardie, "Report of a Visit to ABCC in Japan, June–July 1950 [Copy No. 3]." NAS, 17.

22. Everett Evans and Eugene P. Pendergrass, "Report to Dr. Philip Owen," 30 December 1948, drawer 28, NAS, 8.

ABCC, particularly if they were bilingual, they were no longer expected to have any special influence in the Japanese community.[23]

Despite official statements to the contrary, many Japanese interpreted the ABCC in the early years as an Occupation agency.[24] The ABCC conformed to Occupation policies, and its research concerned a topic of military interest and was thus subject to SCAP censorship.[25] The ABCC was therefore the visible manifestation of the Occupation in Hiroshima and Nagasaki, and the ABCC's relations with Japanese scientists and survivors were constrained and shaped by the fact of Japanese military defeat and Allied occupation. The fine institutional distinctions between the ABCC as a "scientific" agency organized by the National Academy of Sciences and funded by the Atomic Energy Commission, and the Occupation bureaucracy as a military power, were often lost in the day-to-day interactions of Americans and Japanese in Hiroshima and Nagasaki.

Yet the ABCC's American staff members were capable of challenging the Occupation, particularly when Occupation policy threatened a norm of science shared by both Americans and Japanese: the norm of open and free communication. When their Japanese colleagues expressed anger and frustration over Occupation censorship policies, Brues and Henshaw took a personal interest. Encouraged by Tsuzuki, they decided that the publication of Japanese papers on atomic bomb injuries should be made a top priority of the new ABCC.[26]

23. Everett Evans and Eugene P. Pendergrass, "Report to Dr. Philip Owen," 30 December 1948, drawer 28, NAS; also, in same file, Carl Tessmer to Everett Evans, 4 December 1948. Also, for a report on nisei involvement in the ABCC, see Hosokawa (1951). Hosokawa's popular article suggested that the ABCC was still favoring nisei for employment as late as 1951, but internal records of the ABCC indicate that this was not correct. While nisei did make up a significant proportion of the total Allied staff—67 out of 128 in October 1950—this probably reflected nisei interest in working in Japan, rather than ABCC preference. Director Carl Tessmer was interested in bilingual applicants, but many American-born nisei were not fluent in Japanese.

24. Masuo Kodani to Carl Tessmer, 26 August 1948, drawer 19, Genetics—1948 #2, NAS.

25. Censorship generally focused on Japanese findings regarding the biological effects of the bombings. Approximately one hundred Japanese scientific manuscripts were suppressed in the early years of the Occupation, not because they contained any startling information, but because the military and the Atomic Energy Commission did not know how to deal with them. See Monica Braw's broader analysis of American censorship of scientists in Japan (1986).

26. For reports of early meetings with Japanese scientists, see Atomic Bomb Casualty Commission, "Report No. 5, for the Period 22–28 December 1946," 6/A, JVN; Also, James V. Neel, memo re conference with H. C. Kelly of SCAP, 4 February 1948; JVN; and James V. Neel, "Conference on Declassification of Japanese Manuscripts," memo re meeting with Ned Trapnell of the Atomic Energy Commission, 2 February 1948, Drawer 24, NAS.

CENSORSHIP

Occupation policy toward Japanese science went through phases similar to those tracked in other spheres of Occupation influence. The early punitive policies—purges of leading Japanese scientists, for example—were replaced by 1949 with active efforts to build a strong Japanese scientific infrastructure.

One low point in American dealings with Japanese science came early. In November 1945 American soldiers, confused about the possible relevance of cyclotrons to atomic weapons research, dismantled one of Japan's premier research laboratories, the Physical and Chemical Research Institute (Rikagaku Kenkyo-jo, commonly known as Riken). Using blowtorches, dynamite, crowbars, and winches, the soldiers removed operating cyclotrons and dumped them in pieces into Tokyo Bay (Nishina 1947, 145, 167).[27] The incident aggravated the National Academy of Sciences and embarrassed many American scientists (Greenberg 1968, 118–19), but the scientists' concerns had little impact on military policy. In February 1947 the Joint Chiefs of Staff recommended that "all research in Japan, of either a fundamental or applied nature, in the field of atomic energy should be prohibited."[28]

Less dramatic than the incident at Riken was the formal and informal Occupation censorship of all materials dealing with the atomic bombs. As Monica Braw's study suggests, Occupation censors viewed any mention of the bombings as threatening to "public tranquility" and in the interest of this tranquility suppressed poems, fiction, personal testimonials, and scientific reports dealing with the bomb and its effects. For several years, Occupation authorities were reluctant to approve publication of any text that mentioned Hiroshima or Nagasaki, regardless of its tone or content. In 1948, Nagasaki physician Nagai Takashi was not allowed to publish *The Bells of Nagasaki* unless a graphic account of Japanese atrocities in the Philippines, *The Sack of Manila,* was included as an appendix. Nagai's text depicted the dropping of the bombs as acts of "divine providence" and suggested that the use of the weapons was justified. But an American censor explained that any document describing an act of American military aggression must also describe the "Jap military acts that were provocation or motive."[29] Censors feared that publications describing

---

27. Samuel K. Coleman analyzes Occupation policy towards Riken in "Riken from 1945 to 1948: The Reorganization of Japan's Physical and Chemical Research Institute under the American Occupation" (1990); see also Charles Weiner, "Retroactive Saber Rattling?" (1978, 10–12).

28. Joint Chiefs of Staff to Douglas MacArthur, 8 February 1947, 6/C, JVN.

29. Nagai's book was published, with the account of Japanese atrocities in Manila, and became a best-seller in Japan in 1949 (Braw 1986, 103–4).

the bombings could stir resentment against the Americans and the Occupation and reflect poorly on the U.S. decision to use atomic weapons (Braw 1986, 97).

The Occupation also suppressed Japanese scientific publication on the atomic bomb and its effects until 1951, and this became a problem for the ABCC. The form this censorship took was particularly troubling to the Japanese scientific community; rather than explicitly disapproving publication of Japanese work, the bureaucracy would simply issue no decision at all. Japanese scientists who submitted their work, as required, directly to Occupation officials, seeking permission to publish, were informed that scientists in America would review the manuscripts. But most of these manuscripts were never returned, and many scientists never heard what happened to their work.[30]

In a conference with the ABCC team at Riken in December 1946 Japanese scientists complained about these missing manuscripts and about access to publications of other scientists. Aside from occasional literature brought in by visitors, they said, no scientific literature had come into the country since 1940. The "need for bringing Japan back into the sphere of science was urgently expressed."[31] Brues and Henshaw, recognizing that the isolation and disappointment of the Japanese scientific community was not conducive to a good working relationship with the ABCC, wanted to clarify SCAP policy and facilitate Japanese scientific publication. They compiled a list of manuscripts that Japanese investigators reported had been submitted to SCAP for review, and they set out to track down these manuscripts and try to facilitate their publication.

Appealing directly to physicist Harry C. Kelly, deputy chief of the Scientific and Technical Division of the Economic and Scientific Section of SCAP, they asked exactly what Japanese scientists needed to do in order to publish their work. Kelly's response was not entirely satisfactory. He said SCAP had "no objection in principle to the publication of papers on research in any scientific field" and that this policy had been publicly announced the previous month.[32] But Brues and Henshaw were skeptical. "After the story we had received so consistently from the Japanese, it seemed incredible that this statement could have been made publicly. We were assured, however, that it had

30. In 1952, nine of these missing manuscripts were discovered in an office at the ABCC in Hiroshima—a matter of some embarrassment to then–ABCC director Grant Taylor. Unpublished manuscript, GT.

31. Atomic Bomb Casualty Commission, "Report No. 5, for the Period 22–28 December 1946," 6/A, JVN.

32. See Atomic Bomb Casualty Commission, "Report No. 6, for the Period 30 December 1946 to 4 January 1947," 6/A, JVN.

been submitted to the Japanese press and that there was no restraint on the publication of Japanese scientific information."[33]

The problem of the Japanese manuscripts was a "pressing one from the point of view of Japanese-American cooperation," Philip Owen of the National Research Council observed, after Brues and Henshaw began their efforts to track down the missing manuscripts. Owen reported that "Japanese scientific papers are now reduced to a trickle due to this factor of discouragement."[34] An NRC representative, Herman Wigodsky, speaking on behalf of the ABCC, met with Joint Chiefs and AEC officials to discuss the problem. Wigodsky explained the NRC's "immediate desire" that the Japanese manuscripts submitted to the Occupation in 1946 and 1947 be freed for publication. He learned, however, that the manuscripts were caught in a bureaucratic bind that reflected the broader problem of how the United States should handle atomic secrets in the immediate postwar period.[35]

The Japanese manuscripts listed in Brues and Henshaw's 1947 report had been sent to the Army Institute of Pathology, which provided copies to the Joint Chiefs of Staff for review.[36] The Joint Chiefs asked the Atomic Energy Commission to render an opinion as to whether the Japanese should be allowed to publish these reports, but the AEC merely responded that it was "prepared to release the material if there was no military objection." This perturbed the Joint Chiefs, who wanted to be told specifically whether any scientific information contained in the reports was classified or restricted. If the information was classified they would be unable to circulate the manuscripts throughout the War Department in order to ascertain military objections.[37] The 118 Japanese manuscripts listed by Tsuzuki and included in the Brues-Henshaw report were not cleared by SCAP, the Joint Chiefs of Staff, or the AEC until January 1949. Scientists who had submitted their manuscripts to SCAP in 1945 waited almost four years for an answer, apparently because the Americans responsible could not decide how to handle the reports.[38]

33. There were in fact significant restraints on Japanese publication of scientific findings relevant to atomic energy or atomic weapons. See Atomic Bomb Casualty Commission, "Report No. 6, for the Period 30 December 1946 to 4 January 1947," 6/A, JVN.

34. He was citing SCAP's Harry Kelly on this point. Philip Owen to Herman Wigodsky, 12 February 1948, 6/J, JVN.

35. Philip Owen to Herman Wigodsky, 12 February 1948, 6/J, JVN.

36. "Report of the ABCC—January 1947," generally referred to as the Brues-Henshaw Report, NAS.

37. James V. Neel, "Conference on Declassification of Japanese Manuscripts," 2 February 1948, drawer 24, NAS.

38. Atomic Bomb Casualty Commission, "Report No. 6, for the Period 30 December 1946 to

Censorship was not technically the ABCC's concern. But in practice it conflicted with the idealized vision of science that guided ABCC planners and placed a barrier between the Americans involved in the ABCC and their Japanese counterparts. Collaboration and cooperation between Americans and Japanese would be impossible if the Americans were free to publish and the Japanese were not. The point was not that the Japanese publications would contribute to scientific knowledge—many American scientists were skeptical about Japanese ability—but that suppression of reports for political reasons violated scientific norms. Similarly, SCAP's mandate that the ABCC affiliate with the new Japanese NIH, an agency created by the Occupation, isolated the Americans from Japan's most prestigious research institutions and its leading scientists.

THE JAPANESE NIH

Occupation policy required that any group working with the Japanese people work with and through some Japanese agency. An American agency such as the NRC or ABCC could not directly employ Japanese nationals. The Japanese National Institute of Health (JNIH) then, as liaison to the ABCC, formally employed the Japanese nationals hired by the ABCC in Hiroshima and Nagasaki. Colonel Crawford Sams, whose office, the Public Health and Welfare Section of SCAP, had modeled the JNIH on the National Institutes of Health in the United States, was eager in 1947 to promote the new organization. He encouraged the ABCC alliance with the JNIH and promised Neel that it would "in time become an outstanding research and teaching center in Japan, as well as serving many necessary public health functions."[39]

Unfortunately, the JNIH remained "an emphatic Occupation agency" throughout the Occupation. Sams carried out a vigorous purge of Japanese science in the fall of 1945, removing from positions of authority those who had worked with the Japanese military during the war. This included many of Japan's most respected scientists. Because the JNIH was intimately tied to Sams and the Occupation, its scientists were often those who had not been prominent enough to play important roles in the war, or, as the AEC's John Bugher put it, "persons of a relatively inferior calibre."[40]

---

4 January 1947," 6/A, JVN. Also, Braw's examination of the problem from the perspective of both Washington and Tokyo (1986, 120–36).

39. James V. Neel to Crawford F. Sams, "Memorandum: Conference Concerned with the Genetic Program of the Atomic Bomb Casualty Commission," 23 May 1947, 6/J, JVN.

40. John C. Bugher, "Supplement to Report of the Atomic Bomb Casualty Commission Field

While the JNIH had potential, the university system had superior scientists in 1947. The ABCC alliance with the JNIH, however, made it difficult for university scientists to work with the ABCC. The extreme compartmentalization of Japanese bureaucracies meant that persons affiliated with an institution under one ministry could not work under another even as consultants. Kida F., for example, a geneticist with the Kumamoto Medical College, worked with the ABCC briefly in early 1948. He wished to continue his work in Hiroshima but, unwilling to give up his academic affiliation, was forced to resign from the ABCC in March, when the Japanese government abolished a program allowing "part-time service" in other ministries. The JNIH invited Kida to accept a "sub-membership" that would have allowed him to stay with the ABCC, but this plan was not acceptable to the medical college.[41]

Kida's resignation may have been overdetermined. SCAP laboratory consultant Howard Hamlin hypothesized in September 1948 that Kida resigned because he was annoyed by the "interference" of two staff members in the Disease Prevention Bureau of the JNIH. These two staff members were apparently in de facto control of the JNIH—its director Kobayashi R. was, Hamlin observed, a "weak" administrator and had permitted these "petty officials" to make important policy decisions. Hamlin said that Kida sought Kobayashi's guidance on the employment of an assistant for the ABCC genetics program, but one of the "petty officials" "broke into the conversation and stated flatly that such a request could not be granted." Kida resigned because he "did not wish to associate himself" with such an organization. Hamlin added that "it is commonly reported from various sources that 'first class' scientists will not affiliate themselves with the institute until this situation is remedied."[42] Kida's own letter explaining his resignation to ABCC geneticist Masuo Kodani (a Japanese-American) officially blamed the change in policy regarding "part-time service." But later in the letter Kida hinted that reservations about the JNIH were also involved: "As long as NIH remains at the center of the project, I feel that I am not qualified to work on the project."[43]

Kida suggested that the universities, under the Ministry of Education, and the JNIH, under the Welfare Ministry, cooperate with the ABCC on an equal

---

Operations, 27 March to 7 May 1951," drawer 28, Dr. John C. Bugher (AEC) Report on ABCC Visit—March–May 1951, NAS.

41. Some ABCC staff members suspected Kida of sympathy for the hereditary theories of Lysenko, and he also apparently favored ABCC treatment of the survivors; it may be that he would not have been such a welcome addition after all. See Ray Anderson to James V. Neel, 3 November 1948, drawer 19, Genetics—1948 #2, NAS.

42. Howard Hamlin, "Memorandum for the Record," 2 September 1948, 6/F, JVN.

43. Kida F. to Masuo Kodani, 28 March 1948, drawer 19, Genetics—1948 #2, NAS.

basis. Eight years later the ABCC followed this advice, but for the important period during and immediately after the Occupation, the ABCC was burdened with an unproductive institutional alliance dictated by SCAP policy.[44]

In addition, the affiliation with the JNIH, which formally employed all Japanese working for the ABCC, resulted in unequal pay scales. This problem was particularly acute given the rate of inflation and caused considerable hardship for those at the lower end of the scale. One Japanese nurse, in a heartrending appeal to ABCC director Carl Tessmer in 1948, described her and her coworkers' plight in vivid terms. Nurses on the ABCC payroll received four thousand yen per month, she said, while those working beside them but employed by the JNIH received only thirteen hundred yen. "Why are we, who are on the NIH payroll, so poorly paid working in the same commission?" A dress suitable for work cost ten thousand yen—almost eight months' salary for the JNIH nurses—and a new pair of shoes cost three months' salary ("We are too shy to come to office in Japanese clogs"). Moreover, the nurses had to pay almost half their monthly salary—six hundred yen a month—for transportation to the ABCC. "We cannot buy and eat enough nutritious food to maintain our health: it is too expensive for us with the present salary. Already there are some who have lost weight." She added that the nurses had not complained because "silent endurance has been a sad tradition of Japanese women, and we have followed that tradition quite blindly" but "there is a limit to our patience."[45] Tessmer was not unmoved by this appeal. The ABCC staff recognized that the pay scale differential was not fair. But the JNIH, despite repeated appeals to the Japanese Diet, did not have enough funding to match ABCC pay scales.[46]

As one early consulting team noted, the poor relationship to Japanese science was "one of the most disturbing signs in our whole Japanese program." The consultants blamed this directly on the connection to the JNIH.[47] Affiliation with the Educational Ministry would have strengthened the ABCC's ties

---

44. In 1956, Dr. Nakaidzumi Masanori, a retired professor of radiology from Tokyo University, became an associate director of the ABCC. This placed him on an equal level with Maki Hiroshi, the JNIH representative in the ABCC. Kida F. to Masuo Kodani, 28 March 1948, drawer 19, Genetics—1948 #2, NAS.

45. Letter translation attached to Carl Tessmer, "Memo for the Record," 10 July 1948, drawer 11, NIH Japan, NAS.

46. The Hiroshima Medical College in this period paid one-quarter-time interns one thousand yen per month, and half-time interns two thousand yen per month, and full-time interns five thousand yen per month. See Carl Tessmer, "Memo for the Record," 29 June 1948, Drawer 11, NAS.

47. Everett Evans and Eugene P. Pendergrass, "Report to Dr. Philip Owen," 30 December 1948, drawer 28, NAS.

to Japanese science and helped in recruiting Japanese staff. By allying itself with the JNIH, the ABCC cut itself off from access to the top scientific talent in the country.

Yet the alliance with the JNIH promoted Occupation policy. Occupation officials sought to encourage the development of an NIH on the U.S. model, and they saw the JNIH as a vehicle through which democratic values could infiltrate Japanese science.

CONCLUSION

In January 1947, Colonel Sams, commanding officer of the Public Health and Welfare section of the Occupation, called a press conference to announce that the Americans were launching a medical study of the atomic bomb survivors. In February Brues and Henshaw returned to Washington, leaving Neel and Block in Japan.[48] By March Neel had the necessary approvals from SCAP. He received two jeeps through the Chugoko Military Government Region of the Occupation. And Fred Snell, a physician with the Occupation, arrived to help out.[49]

That spring Neel and Snell hired the first three Japanese employees of the ABCC: a physician-interpreter, a physician-technician, and a secretary.[50] A few patients visited the laboratory in Kure, and Neel and Snell began collecting blood samples for the proposed hematological study. They had only two small rooms at the Red Cross Hospital, accommodations Neel judged "none too adequate." Yet, he said, "to take more space in a city as devastated as Hiroshima would be unfair." [51]

Hiroshima in late April 1947 was still in chaos. Residents had cleared the streets, and many had erected small shacks of wood, plastered mud, or bamboo. These shacks had no plumbing or heat. Food was still a serious problem.

48. Atomic Bomb Casualty Commission, "Interim Report No. 1, Covering the Period 8 January 1947 through 20 January 1947" and "Report No. 6, for the Period 30 December 1946 to 4 January 1947," 6/A, JVN.

49. Atomic Bomb Casualty Commission, "Interim Report No. 6, Covering the Period 14 April to 28 April 1947," 6/A, JVN.

50. One Japanese employee remembers Japanese employment as beginning slightly later. Moriyama Isao, secretary of the ABCC from 1948 until 1984, has recalled that the first Japanese applicants for employment were interviewed in Kure in November 1947. Moriyama was hired on 5 January 1948. He began his career at the ABCC as a "contactor" of survivors in Hiroshima (Moriyama 1988, 43–44). Also see Atomic Bomb Casualty Commission, "Interim Report No. 5, Covering the Period from 14 March 1947 to 13 April 1947," 6/A, JVN.

51. Atomic Bomb Casualty Commission, "Interim Report No. 6, Covering the Period 14 April to 28 April 1947," 6/A, JVN.

Wheat, radishes, and other vegetables were growing in every vacant patch, presumably a result of efforts to compensate for food shortages. Neel, on a walking tour of the city, noticed that the human shadows against the burned asphalt of the bridges in Hiroshima had disappeared. But the "stone lanterns on the bridge near the parade grounds are still asymmetrically displaced and the heat blasted granite of the Gokuku Shrine still contrasts strikingly with its polished protected surface." The population of Hiroshima was 210,000, up from an estimated 170,000 in the fall of 1945. People who had left the city after the bombings were coming back, and new residents were coming in at a brisk rate.[52]

In late May of 1947, Neel was summoned to Washington for a meeting of the Committee on Atomic Casualties, the NRC oversight committee of the new ABCC. This left Snell to run the program in Japan. That August the National Research Council received the "long-awaited letter of intent" from the Atomic Energy Commission, promising to fund a large-scale, long-term biomedical study of the survivors. Neel, then still in Washington, planned to return to Japan in September, spend six months getting the genetics study going, then return to the United States to take up a faculty position waiting for him at the University of Michigan.[53]

Snell, left on his own in Hiroshima to direct the small ABCC staff, enlisted the local police to round up survivors and began scheduling twenty appointments a day to draw blood. Unfortunately, most of those called in had no transportation to the Red Cross Hospital. Snell had only two jeeps, and one of them spent most of the summer in Kobe, northeast of Hiroshima, supposedly being repaired by military mechanics. These transportation difficulties were "producing a real resentment" among the survivors, who had to "endure the hardships of civil transportation, made worse by the fact that most of them must travel far." Snell was also trying to negotiate more reasonable quarters for the ABCC. He wanted the local library building, not then in use, but the mayor of Hiroshima was determined to restore it as a library.[54]

Thus the Atomic Bomb Casualty Commission began its work. The local police were serving as "contactors." The on-site organization consisted of one American staff member and three Japanese crowded into two rooms at the Red Cross Hospital in still-devastated Hiroshima. The process of hiring Japanese employees, who would do much of the primary work of the ABCC, had begun.

52. Atomic Bomb Casualty Commission, "Interim Report No. 6, Covering the Period 14 April to 28 April 1947," 6/A, JVN.

53. James V. Neel to Harry Johnson, 7 August 1947, 6/J, JVN.

54. Fred Snell to James V. Neel, 28 August 1947, 6/B, JVN.

CHAPTER FOUR

The Genetics Study

I n October 1947 a brief essay appeared in *Science* under the headline "Genetic Effects of the Atomic Bombs in Hiroshima and Nagasaki." This essay was a disclaimer. Prepared by the National Research Council committee that would help direct the ABCC's genetics study, it said that an American study of the genetic effects of the radiation resulting from the atomic bombings would be undertaken, but with the understanding that statistically significant results were unlikely. Unfortunately, the authors noted, the negative findings might be popularly interpreted as meaning that genetic effects had not occurred. This would not be a legitimate conclusion, they stressed, since there was every reason to infer that heritable changes had been produced at Hiroshima and Nagasaki. However, for a variety of reasons, these genetic effects would probably not be demonstrated in this or any other study (Genetics Conference, Committee on Atomic Casualties 1947). With this modest statement, 57 the genetics project of the Atomic Bomb Casualty Commission was formally launched.

That the published statement could easily be misunderstood is suggested by the fact that a scientist of the stature of Detlev Bronk, biophysicist at the University of Pennsylvania and later president of the National Academy of Sciences, could, slightly more than two years later, recall that the essay had concluded that the ABCC's "investigation will reveal increased genetic effects."[1] The NRC committee had in fact agreed that something that could be called mutations, changes at the level of the hereditary material, would be present. But it also indicated that it did not expect these changes to be manifested frequently enough in the children of survivors to be statistically significant in the first generation born after the bombings.

The announcement in *Science* was intended to inform the scientific community that spectacular results were unlikely.[2] It was a way of keeping expectations within reason. From the AEC's perspective, supporting the work on the survivors was necessary regardless of the expected outcome. As an AEC committee concluded in 1948, "While the long-range scientific value of the program was perhaps questionable, the demand by the public for information was so great that it would be impossible to fail to support the study without making the Atomic Energy Commission appear derelict in its responsibilities."[3] Neel told a colleague in 1947 that the "consensus of opinion was that while from the laboratory standpoint this was a highly unsatisfactory 'experiment' nevertheless because of military, industrial and certain scientific considerations a rather large-scale program seemed indicated."[4]

By 1947, radiation's biological effects on cellular processes were well known. Radiation was used therapeutically to kill rapidly dividing cancer cells; it was a popular research tool in genetics, known to cause heritable mutations; and it had been linked to skin cancers (in X-ray workers), leukemia (in radiologists), cataracts (in cyclotron workers), lung cancer (in those who inhaled radioactive particles), and bone cancers (in those who ingested radioactive materials).[5]

---

1. Bronk made this comment in the context of a discussion about the value of the ABCC as a whole. See Advisory Committee for Biology and Medicine, Atomic Energy Commission (hereafter ACBM), "Minutes, 18th Meeting," 20 October 1949, RG 326, box 3218, ACBM, DOE, 45.

2. James V. Neel, interview by author, Ann Arbor, Michigan, 4 February 1988.

3. ACBM, "Minutes," 11 September 1948, DOE.

4. James V. Neel to Harry Johnson, 29 June 1947, 6/J, JVN.

5. Many of these connections were established in the late 1930s and 1940s, though tumors in mice were produced as early as 1910 and the skin cancer effect was recognized by 1900. For a review of the literature available at the time, see A. Glücksman, "Tumour Induction by Penetrating

For geneticists, radiation was a technology for the manipulation of the hereditary material. They were not interested in radiation per se but in genes, and they were professionally and intellectually isolated from biologists whose primary interest was in the biological effects of radiation itself, rather than the manipulation of any particular cellular entity. "Many workers studying the biological effects of radiation do not adequately appreciate the very considerable contributions which have made to this subject by geneticists," said D. E. Lea in the preface to his 1947 summary of radiation research. He went on to explain such terms as zygote, homologous, linkage, and genotype to his readers (Lea 1947, 126–32), presumably radiation biologists unfamiliar with the technical terminology of genetics.

Many radiation biologists believed that there was a threshold dose—a dose below which no biological effects would occur—for the somatic effects of radiation. Debates about a possible threshold dose for radiation have characterized radiation biology for most of its history. This "bitter scientific dispute" is indeed not entirely resolved (see Hall 1984, 31–32). But the idea of a threshold dose never took hold among geneticists. Genetic effects were believed to be linear, that is, to occur in exact proportion to radiation dose: If 100 rads caused six hundred mutations in a given population, 1 rad would cause six. As one biologist explained in 1952, genetic mutations were "classic examples of the non-threshold type of reaction."[6] Unlike somatic effects, which would only appear after a certain level of radiation exposure was reached, genetic effects should therefore occur even if exposure levels were low—as they were, relatively, in the survivors.

While the ABCC was planned to conduct a general study of the "biological effects of atomic radiation" on the survivors themselves, it quickly became, primarily, a study of the genetic effects of radiation on their offspring. The genetics work controlled most of the budget, attracted most of the public attention, and appeared to be the "most important" aspect of the ABCC's work. This was partly a consequence of the scientific conclusions about the threshold dose. But it was also a consequence of the perception, within the NRC and

---

Radiations" in A. Haddow, ed., *Biological Hazards of Atomic Energy* (1952, 87–92, esp. 88–89). See also D. E. Lea, *Actions of Radiations on Living Cells* (1947). Lea's summary of the status of radiation biology in 1947 is highly informative and includes a bibliography that suggests the state of the field. He concludes that radiation effects are mediated through chromosomes, since cells degenerate only at division: "This result is consistent with the idea that the cause of degeneration is some change in the chromosomes" (1947, 343). For an overview of the leukemia data then available, see Henshaw (1944b); see also Shields Warren (1942a). For radiation cataract literature, see G. E. DeSchweinits and B. F. Baer (1932) and P. J. Leinfelder and H. D. Kerr (1936).

6. F. G. Spear "The Biological Response to Penetrating Radiations," in Haddow (1952), 1–6, esp. 2.

beyond, that genetic effects were more threatening or frightening than effects on the survivors themselves.

One goal of the ABCC genetics study was to reassure the public that radiation from the bombs had not produced a generation of genetic monsters. Participants did not genuinely expect "Godzilla," geneticist William J. Schull has noted; the larger public, however, apparently did (Weart 1988; Boyer 1985), and the scientific community could not entirely, unequivocally rule out such a possibility, given the available experimental evidence from laboratory organisms.[7]

Public expectation of radiation monsters at Hiroshima and Nagasaki was thus an important impetus for the ABCC genetics studies. When plans for an American study of the survivors were announced in Japan in January 1947, "about the first question to come up was the matter of possible genetic effects."[8] Press coverage of the ABCC often focused on the importance of the genetics study. As an article in *Life* magazine put it in 1949, the atomic bomb's "toll in human destruction" was not limited to those present in the bombed cities. It could be "carried down through successive generations to bring death and disfigurement to the descendants of the survivors." The study of the genetic effects of radiation was therefore "the most important and difficult of the ABCC's tasks."[9] While leukemia and radiation cataracts had already been demonstrated to be possible consequences of human exposure to radiation, neither problem attracted as much public interest as did genetic effects. Blood studies were begun almost immediately, but when the early organizers of the ABCC listed priorities they often rated genetic effects and keloids as most important. And keloids, the painful scar tissue common after burns in some racial groups, ceased to be of interest when they were shown to be unrelated to radiation effects.[10] The possible genetic effects were discussed in most of

7. William J. Schull, interview by author, Houston, Texas, 16 December 1988.

8. Neel described his comments in a letter to Austin Brues and Paul Henshaw, 31 January 1947, 6/J, JVN.

9. *Life,* "The A-bomb's Children," 12 December 1949.

10. At first the Americans suspected that the severity of the keloids might be a consequence of radiation exposure, but when it began to appear that keloids were not a specific consequence of radiation exposure, ABCC interest diminished. For early speculations on the importance of keloids, see Melvin A. Block to Austin Brues and Paul Henshaw, "Considerations in the Study of Burn Sequelae in Atomic Bomb Survivors," 30 December 1946, drawer 29, no file, NAS. Many Americans persisted in the belief that keloids were a consequence of improper burn therapy. "Our observations lead us to conclude that the development of keloids in the patients who were burned as a result of the thermal effect of the bomb was due largely to exceedingly bad burn therapy. We would conclude that radiation does not seem to assume the importance in the keloid problem that

the planning sessions and public forums in which the ABCC was framed and were a major focus of the research plans.

The genetics study dominated the early years of the ABCC, both financially and intellectually. It became the pillar upon which other activities of the ABCC rested. Much of the commission's overhead cost was routinely charged to the genetics study (a practice administrators came to regret, for budgetary reasons, when the morphological genetics project was terminated in 1954).[11] The genetics project was therefore central to the ABCC's operations, both in terms of its public impact and its internal management.

RADIATION MUTAGENESIS

Scientific interest in the biological effects of radiation began shortly after the discovery of X rays and radium at the turn of the century. Pathological effects of radiation were dramatically documented by 1908, when Charles Allen Porter described more than fifty cases of radiation poisoning to a meeting of the American Roentgen Ray Society (Brecher and Brecher 1969, 165; cited in Caufield 1989, 13). However, despite such evidence of danger, inadequate protection standards for the use of medical X-ray equipment persisted, and both patients and physicians continued to suffer radiation burns and radiation-induced cancers for several decades. Radium even gained a reputation as a health tonic and was widely distributed in patent medicines. In the 1940s, radioactive mines in Colorado began operations as health resorts (some continuing their operations into the 1980s) (Caufield 1989, 27–28). Radiation was a known health hazard, poorly understood and under limited legislative control. And, to a disturbing degree, what protection standards were implemented were based not on scientific data but on technological or professional considerations.[12]

---

it has received in earlier reports." Everett Evans and Eugene P. Pendergrass, "Report to Philip Owen," 30 December 1948, drawer 28, NAS, 9.

11. "We are repeatedly confronted with a piece of logic from the Bureau of the Budget. It runs as follows: if x% of the previous budget was chargeable to the genetics program x% should now be cancelled or released for other use now that genetics is virtually eliminated. We are hoist with our own petard inasmuch as we had charged so much of the fixed overhead to genetics because the latter was a favored child." Keith Cannan to Robert Holmes, 10 November 1954, drawer 2, NAS.

12. See Gilbert F. Whittemore's exploration (1986) of how radiation protection standards depended more on the perceived impact of new regulations on industry than on scientific data. See also Barton Hacker, *The Dragon's Tail: Radiation Safety in the Manhattan Project, 1942–1946* (1987); Catherine Caufield, *Multiple Exposures: Chronicles of the Radiation Age* (1989); and John Gofman, *Radiation and Human Health* (1981).

For biologists, X irradiation quickly became a tool for the manipulation of biological processes,[13] but until 1920 no researcher carried out specifically genetic experiments with radiation. That year Nadson and Philippov used radiation to produce what they surmised were new stable races of fungi (1925, 93). Early attempts to induce mutations in mice in the 1920s were unsuccessful, but work with the common fruit fly, *Drosophila melanogaster,* was more productive. By 1921, James Watt Mavor, a professor at Union College in Schenectady, New York, was able to identify radiation-induced X-chromosome loss, nondisjunction, and effects on crossing-over in *Drosophila.*[14] In 1925 H. J. Muller began his Nobel Prize–winning work on the effects of X rays on crossing-over and chromosome breakage in *Drosophila melanogaster* (Carlson 1981).

Muller's ingenious approach allowed him to detect sex-linked recessive lethals induced by exposure to radiation. He showed that X rays produced heritable mutations, and that these mutations were of the same types as known spontaneous mutations (Muller 1927). Muller's publications inspired other investigators, who by 1934 had carried out radiation studies on four species of *Drosophila* (*D. melanogaster, D. pseudoobscura, D. funebris,* and *D. virilis*); various plants, including wheat, oats, barley, rye, maize, cotton, vetch, tomatoes, and nicotiana; and one mammal, *Mus musculus,* the house mouse (Timofeef-Ressovsky 1934, 418–19). These studies suggested the general effectiveness of radiation in inducing heritable mutations of various kinds in all species tested.

These investigators were interested in radiation primarily as a biological probe, a means of artificially creating a trait that could be used to reveal the mechanisms of heredity. Muller recognized that his work had implications for

---

13. Bardeen studied the pathological effects of radiation on biological tissue as early as 1906. Paula Hertwig also began exploring X-ray and radium effects on animal germ cells, plants, and chromosomes (Timofeef-Ressovsky 1934)

14. Mavor published a report on this work in *Science* in 1921. A more comprehensive account appeared in the *Journal of Experimental Zoology* in 1924. Historians have not ranked Mavor with Muller and Stadler as an early discoverer of the genetic effects of radiation, but the reasons for his relative obscurity are unclear. Mavor himself, in his 1952 biology textbook *General Biology,* took full credit for the discovery and mentioned Muller only in a footnote, as one of the "other investigators" who confirmed his findings "several years later" (Mavor 1952, 653). Muller won the Nobel Prize in 1946 for his discovery of the genetic effects of radiation—which Mavor must have known—so Mavor's 1952 assessment of Muller was perhaps intentionally provocative. Ernest B. Hook (1986) explores the question of Mavor's contributions and his interaction with geneticists in the 1920s in his "James Watt Mavor (1883–1963): A Forgotten Discoverer of Radiation Effects on Heredity."

radiation protection standards, noting in his first paper on the subject that "it would seem incumbent for medical practice to be modified accordingly." His goal, however, was not to understand radiation but to understand heredity (1927). The issues addressed in early research on radiation mutagenesis therefore reflected debates in genetics and theories of evolutionary change.[15]

These early studies suggested that different loci mutated at different rates and that some forms were more likely to appear than others. Thus, for example, the various white-eye morphs in *D. melanogaster* differed in their sensitivity to radiation mutagenesis. The darker shades of white mutated more frequently under the same radiation dosage than the lighter shades. These differences in mutability—between individuals of the same species, from locus to locus, and, presumably, from species to species—suggested some of the difficulties of extrapolating from the animal and plant studies to human populations.

Human beings have been irradiated occupationally, medically, and as a consequence of exposure to nuclear weapons and nuclear power accidents. But such irradiation has not usually affected a large number of people, and except in cases of therapeutic medical irradiation, doses have been unknown or difficult to estimate. And attempts to measure the genetic effects of irradiation face an additional complication: Any second-generation abnormalities that might be taken as signs of mutation promise to be indistinguishable in kind from the birth anomalies known to occur in any population. While a higher rate of such abnormalities might be detectable, no single malformation could be taken as a specific sign of radiation-induced changes in the hereditary material.

Before the war the genetic effects of radiation did not seem important to some scientists on the U.S. Advisory Committee on X-ray and Radium Protection (after 1946, the National Council on Radiation Protection). As Gilbert Whittemore's work has demonstrated, genetic effects were a low priority for Giannochio Failla, a Columbia University physicist who served on the committee. Failla felt that radiation tolerance doses should be set based only on

15. The signs recognized in *Drosophila* as indicative of radiation effects included changes in the sex ratio (Muller's work) and variations (minor and major) in eye color, wing shape, wing venation, bristles, and eye shape. Plant species displayed variegated coloration and form, stunting, and reduced fertility. "Fitness" parameters frequently used in animal studies included survival, fertility, and life span, all characterized by relatively low levels of heritability. Other measures used in the early radiation studies included skeletal abnormalities and learning ability. Reductions in mouse litter sizes were confirmed early in radiation studies, as was the presence of "semi-sterility," the term widely used in mammalian genetics to refer to a condition in which litters are about half the normal size. The term of course does not literally make sense—since one cannot be half sterile any more than one can be half pregnant—but it is well entrenched in the literature.

damage to individuals and not on possible damage to their offspring. The somatic risks of radiation were even more poorly quantified in the late 1930s than the genetic risks, so Failla's concerns reflected not the state of scientific knowledge but the practical problems of managing the use of radiation in medical therapy (Whittemore 1986, 135–39).

In 1941, a serious plan for a study of the genetic effects of radiation was proposed by Paul S. Henshaw, senior radiobiologist at the U.S. Public Health Service (later to play an important role in the Atomic Bomb Casualty Commission). Henshaw, who had prepared a literature search on the biological effects of radiation for the Manhattan Project that year, wanted to survey radiologists and others occupationally exposed to assess the effects of that exposure on their children (Henshaw 1944b). In October 1941, he appealed for financial support to Lauriston Taylor, a physicist at the Bureau of Standards who also served as chair of the Committee on X-Ray and Radium Protection (Whittemore 1986).

Henshaw's plan called for a five-page survey to be sent to a thousand persons exposed to radiation through their employment (either as radiologists or technicians assisting radiologists) and to a thousand persons engaged in medical work that did not involve exposure to radiation. Respondents would be asked to estimate their exposure level and to describe any abnormalities in their children. Respondents would also be asked to describe abnormalities present in their parents or siblings so that family history could be taken into account (Whittemore 1986, 184–86).

Some of the radiation protection committee members objected to the study, concerned that the questions would alarm technicians. "It might stir up trouble between a technician and an employer," one NCRP member objected. Another said the questionnaire was "likely to create a minor furor" among employees of radiologists. Their proposed solution was to restrict the survey to "management." Others objected that the study would provide unwarranted reassurance about radiation effects. "I am seriously afraid that the undertaking of this program by the National Cancer Institute would be taken as a formal decision on the part of the Institute that the question of roentgen injury to man's heredity can be answered by such a questionnaire," objected one (Whittemore 1986, 188–89).

Committee member R. R. Newell, who strongly favored incorporating estimates of genetic injury into radiation protection standards and who was sympathetic to genetic concerns, also questioned the Henshaw proposal. Newell said the study would not give "even a probable answer" to the question of genetic effects. If abnormalities were found, he said, they would not necessarily be the product of radiation mutagenesis. If they were not found, it would not settle the issue in any way. "I for one am going to be very skeptical about the degree of reassurance that will provide." [16] Because of these objections, and the consequent lack of funding, Henshaw's proposed survey was never carried out. Not until 1955 was a survey of radiologists regarding their offspring completed. [17]

Meanwhile Donald R. Charles began secret radiation work with mice in 1943, under contract to the Manhattan Engineering District, as part of the Rochester Atomic Energy Project. Scientists developing atomic weapons needed to assess the potential biological effects of radiation on personnel exposed at Oak Ridge and other locations. Charles's work therefore focused on the effects of chronic, low-level irradiation, rather than single high doses such as those experienced by survivors at Hiroshima and Nagasaki. Yet it influenced Neel's construction of the genetics study, partly because Charles's goal was to answer a "fundamental question" that was closely related to the question driving the ABCC studies: "What is the probability that a person who has intercepted a certain small amount of radiation daily for a certain number of days will, as a result, transmit to a subsequent child a genetic change which will be harmful to the child?" [18] Charles accepted the notion that mice could provide an answer to this question about human exposure.

Both Snell (1935) and Hertwig (1939) had used the house mouse in radiation studies, but neither with such keen attention to the implications of their work for human populations. Charles specifically constructed his experiments with mice to explore questions about human exposure to chronic, low-level irradiation. Experimental animals, he said, should be "examined and tested very thoroughly for those types of genetic change which would be significant in human populations" (Charles et al. 1960, 7). He identified changes significant in human populations as reduced fertility, disturbed sex ratio, and abnormalities of structure or function. Charles included reductions in "fitness"

16. Whittemore explores the reaction to the Henshaw proposal (Whittemore 1986, 180–96). See particularly pages 191–92, on Newell's response; the letter cited is Newell to Taylor, 12 January 1942, Box 31, NCRP—1942, papers of Lauriston Taylor. See also, Newell to Paul S. Henshaw, 9 February 1942, Box 31, NCRP—1942, papers of Lauriston Taylor, cited by Whittemore on page 195.

17. This is the study by Stanley Macht and Philip S. Lawrence, "National Survey of Congenital Malformations Resulting from Exposure to Roentgen Radiation" (1955). The survey was made in 1951 with support from the National Institutes of Health. I explore the results in more detail in my chapter 11, on the publication of the genetics data in 1956.

18. This question is posed in the 1960 report on Charles's work. The report was written up, with Don Charles as senior author, by three coworkers, Joseph A. Tihen, Eileen M. Otis, and Arnold B. Grobman. It was never properly published but appeared as an AEC Research and Development Report titled *Genetic Effects of Chronic X-Irradiation Exposure in Mice* (Charles et al. 1960). Grobman had already published a popular account of the mouse studies in 1951.

parameters, such as life span, strength, weight and length, and total number of offspring, as "abnormalities of structure or function."

Charles and his coworkers, including Joseph A. Tihen, Eileen M. Otis, and Arnold B. Grobman, irradiated male mice at five levels (0.1 roentgen unit per day, 0.5 roentgen unit per day, 1 roentgen unit per day, 10 roentgen units per day, and 0 roentgen units per day, as controls) and studied their offspring to detect genetic changes.[19] These mice were bred with nonirradiated females, producing 13,838 offspring over a period of seven years, from 1943 to 1950 (Charles et al. 1960, 8–9).

The female offspring of irradiated males were allowed to produce four litters, then killed and examined. If these females were found to have any abnormalities, their offspring (the F2) were bred to test for heritability of the trait, which might appear in the next generation. The male offspring of irradiated males were bred only if they had a visible abnormality. Abnormalities detected in males only at autopsy could not be tested for heritability, because the affected mice had not been bred (Charles et al. 1960, 10).

As the experiment was set up, each exposed male in the first generation would be represented by approximately 1,300 mice in the F2. Presumably genetic effects would be detectable in such a large population (Charles et al. 1960, 10). Charles was in fact able to detect small but consistent signs of radiation effects, particularly in "total mutations" and litter size. The total number of abnormal offspring ranged from 0.152 percent for the controls, to 1.333 percent for mice subjected to chronic irradiation of 10 roentgen units (about 10 rads) per day (Charles et al. 1960, table 68, 342). Other parameters revealed no clear trends. Sex-ratio results were inconclusive, partly because data on sex

19. Female somatic effects are more difficult to isolate from genetic effects, since the "maternal environment" might have an impact on some of the parameters being studied (stillbirth, for example). In addition, chronic irradiation of females during pregnancy could cause damage to their developing fetuses which might be indistinguishable from genetic effects. For simplicity, then, most investigators irradiated only males.

Charles was using roentgen units, which were at the time the common unit used to measure X rays. Roentgen units are not exactly equivalent to rads, but were used interchangeably in the early years of radiation biology, since one roentgen is equivalent to the absorption of 93 ergs per gram of tissue at the body surface and one rad is equivalent to the absorption of 100 ergs per gram of tissue. The difference was not particularly important, considering all the other complexities of these earlier studies. The erg was the unit of energy commonly used in radiation biology to characterize the amount delivered to a particular mass of tissue. Any other unit could be used—calories, joules, BTUs, electron volts, etc.—and a more recent measurement, the Gray, does use the joule rather than the erg. One gray refers to the absorption of one joule of energy per kilogram of tissue. As things work out, 1 Gray is equal to 100 rads, simplifying conversion. Rads were the dominant measurement used in the scientific literature until the 1980s, when Grays became increasingly more popular. See Gofman (1981, 42–53).

ratio at birth (not postweaning) was lost and no longer available when the data were analyzed for publication (Charles et al. 1960, 346). Postweaning mortality in the offspring of exposed males seemed to be higher, but the data were not clear. Litter size differences were significant. Charles found that paternal irradiation was clearly correlated with reductions in litter size and emphasized that the finding was based on "test-proven results" rather than "any inferences from the size of litters." (Charles had sacrificed some of his females at various stages of pregnancy to count embryos.)[20]

As a part of the Manhattan Project, Charles's work was secret during the war. For a variety of reasons, he did not publish his results before his death in 1955. In 1960, several colleagues pulled together his notes and prepared a report for publication by the AEC. By this time, William Russell at Oak Ridge had surpassed Charles's work with mouse irradiation.[21] Charles's studies were therefore not particularly influential in radiation genetics or mouse genetics. But they did have an impact on James Neel, who had access to Charles's secret reports in 1946 and sympathized with Charles's approach, particularly his emphasis on "fitness" variables such as growth and weight.

Neel was impressed that Charles chose to sacrifice females to count embryos, a decision which permitted him to show a clear correlation between paternal irradiation and reduced litter size. He was sympathetic to Charles's basic interest in signs of mutation that could be expected to be important in human populations. Neel also knew Charles personally—they had been at Rochester together—and this may have increased his faith in Charles's results.[22]

NEEL AND HIS INVOLVEMENT IN THE ABCC

When the ABCC project began, James Neel was a relative newcomer to human genetics, thirty-six years old and trained in *Drosophila* genetics.[23] A student of

20. Charles estimated from his findings an induced mutation rate of 3.85×10^{-5} per roentgen unit (Charles et al. 1960, 349).

21. See Russell (1956).

22. Neel, interview with author, 4 February 1988, Ann Arbor, Michigan.

23. Neel's *Drosophila* publications include: "Phenotypic Variability in Mutant Characters of *Drosophila funebris*" (1937); "The Interrelations of Temperature, Body Size and Character Expression in *Drosophila melanogaster*" (1940); "Studies on the Interaction of Mutations Affecting the Chaetae of *Drosophila melanogaster:* I. The Interaction of Hairy, Polychaetoid and Hairy Wing" (1941b); "A Relation Between Larval Nutrition and the Frequency of Crossing Over in the Third Chromosome of *Drosophila melanogaster*" (1941a); "A Study of a Case of High Mutation Rate in *Drosophila melanogaster*" (1942b); "The Polymorph Mutant of *Drosophila malanogaster*"

Curt Stern (shortly after Stern immigrated from the Kaiser Wilhelm Institute in Berlin), Neel graduated with a Ph.D. in genetics from the University of Rochester in 1939. He immediately accepted a position as an instructor in zoology at Dartmouth College, where he stayed for two years. He attended the International Congress of Genetics in Great Britain in the summer of 1939, and there met Charles Cotterman, a geneticist then at the University of Michigan, who was also interested in human genetics and who would later be, with Neel, a key player in the early development of the American Society of Human Genetics.[24]

In 1941, as a National Research Council fellow, Neel went to Columbia to work with Theodosius Dobzhansky and Leslie Dunn. He was still working with *Drosophila* genetics but already planning to shift his focus to human genetics.[25] Neel's commitment to human genetics was unusual considering the status of the field in 1941. Academic jobs for human geneticists were rare to nonexistent, the field was tainted by the excesses of the American eugenics movement, and the methodological problems posed by the study of human populations appeared to many geneticists as nearly insurmountable. Neel interrupted a promising career in fly genetics to attend medical school at the University of Rochester because he felt that a human geneticist needed to be trained as a physician as well as a geneticist. His plan was risky when there was no clear path for an academic human geneticist to follow (McKusick 1975). Characteristically, however, Neel was content to create his own path.

His first project in human genetics, published in 1943, was a modest assessment of the inheritance of red hair, based on a perusal of records at the eugenics office at Cold Spring Harbor, New York (Neel 1943a). While many of his writings are informed by his concerns about the future of human evolution, Neel was never a supporter of the American eugenics movement. The eugenics records contained some useful pedigrees, but Neel spent much of his professional career working to establish human genetics as a serious scientific discipline untainted by the ideology of mainline eugenics.[26]

(1942a); and "Studies on the Interaction of Mutations Affecting the Chaetae of *Drosophila melanogaster:* II. The Relation of Character Expression to Size in Flies Homozygous for Polychaetoid, Hairy, Hairy Wing and the Combinations of These Factors" (1943b).

24. William J. Schull provides some personal details on Neel's early life, in "Scientist, Journalist, Orchidist—Will the Real James V. Neel Please Stand Up!" (Schull 1986). On the American Society of Human Genetics, see Ms. Coll. 49, American Society of Human Genetics (particularly Herluf Strandskov's scrapbook), APS.

25. Curt Stern to Lee Dice, 5 November 1945, #4, Neel, papers of Curt Stern, APS.

26. Genetics as a whole, however, has been and continues to be shaped by eugenic conceptions, and in many ways Neel qualifies in the present as a "reform eugenicist." See particularly his popular accounts of the relevance of genetics to human affairs, including "Lessons from a 'Primi-

He began serious work on thalassemia while still a medical student, and published reports on thalassemia in 1944 and 1945 (Neel and Valentine 1944; Valentine and Neel 1945). He also published a paper on inherited cataract (Lutman and Neel 1945). By 1945 he had completed his residency for his medical degree at Strong Memorial Hospital in Rochester. The young geneticist had extraordinary self-confidence, even when dealing with those who might be presumed to be his scientific superiors. His mentor Stern reported that he "occasionally gave the impression to some that he was inclined toward too high a self-esteem," but "this was never serious and in any case he seems to have outgrown any such tendency."[27]

In the spring of 1946 he accepted an appointment as assistant professor of genetics at the University of Michigan. By the following summer—after only one semester in Ann Arbor—he was drafted by the Army and assigned to Cushing General Hospital in Framingham, Massachusetts. Neel's "friends and acquaintances at the Manhattan Project put in an official request for services" as early as March 1946, realizing that Neel would be useful in the project's work on radiation genetics. "One would imagine that such a request would have a fairly high priority," Neel complained to Stern in the summer of 1946. "One would apparently be wrong. Because here I am assigned to *Orthopedic Surgery*."[28] While a general hospital might usually provide many opportunities to learn, Neel noted, this particular hospital was in the process of being transferred by the Army to the Veterans Administration. Neel was therefore buried under the "voluminous paper work" that accompanied such a bureaucratic transfer. He foresaw nothing more exotic in the future than eight weeks of training at a medical field school in Texas.[29] But as Neel prepared to leave for hospital duty in Texas, the Army's instructions suddenly changed. He was ordered instead to Washington to help plan the genetics studies in Japan, and by November he was on his way to Tokyo.

Neel generally explains his involvement in the genetics study by telling this story of his military service. The work in Japan was far more interesting and productive than almost any other possible Army assignment, so he naturally welcomed the opportunity. In addition, as he learned more about Japanese marital patterns he became interested in the possibility of conducting consanguinity studies in Japan. High rates of cousin-marriage provided an ideal test population for the genetic effects of consanguinity. Neel could justify studying this

tive' People: Do Recent Data concerning South American Indians Have Relevance to Problems of Highly Civilized Communities?" (1970).

27. Curt Stern to Lee Dice, 5 November 1945, #4, Neel, papers of Curt Stern, APS.

28. James V. Neel to Curt Stern, 19 August 1946, #4, Neel, papers of Curt Stern, APS.

29. James V. Neel to Curt Stern, 5 November 1946, #4, Neel, papers of Curt Stern, APS.

parameter on ABCC time because it affected the radiation studies: he would have to compensate for differing rates of consanguinity in the exposed and the control populations. In Nagasaki, the bomb's hypocenter was in a heavily Catholic neighborhood. Urakami Cathedral, the largest Roman Catholic church in Japan, was destroyed by the bomb. Catholicism prohibited cousin-marriage, while in Buddhist families cousin-marriages, even between first cousins, were common. Neel would have to study cousin-marriage in order to understand its differing impact on exposed and nonexposed populations. The possibility of such a study was a strong incentive to continue his participation, and the consanguinity studies were later the subject of a much-respected text (with William J. Schull) on the subject (Schull and Neel 1965).

Neel also knew that the genetics study would collect considerable information about the rate of malformation in the Japanese population in general. This was a little-studied population. Population studies could provide clues to historical migrations (Kevles 1986, 164–237; Harris 1959). There were therefore tangential scientific projects that sparked Neel's interest in the Hiroshima-Nagasaki project, beyond its potential for revealing the genetic effects of radiation.

In August of 1947 Neel accepted the title of "interim director" of the ABCC. He officially became a civilian on 7 September 1947, after the NRC requested his immediate release from the Army contingent on his return to Japan to work with the bomb survivors. Neel planned to spend about six months in Japan setting up the study, then to return to academic life in the United States. He looked forward to this and told a colleague in September 1947 that it would be "fine to climb up in the ivy-covered ivory tower of academic life for which it seems I am much more suited by temperament than the work in Japan."[30]

But the six-month commitment to Japan turned into a relationship that has lasted more than forty-five years, and while Neel did return to his academic appointment at the University of Michigan, he never left the Japan project entirely behind. He has remained involved in the work in Japan off and on, as a consultant or participant, up to the present. The motivation for this continued participation seems to be a blend of social commitment and stubbornness.

THE GENETICS CONFERENCE

Neel formally unveiled his detailed plans for the ABCC genetics project at a special "genetics conference" in Washington in June of 1947. H. J. Muller was

---

30. James V. Neel to Fred Snell, 8 September 1947, JVN.

among the participants. Also present were Donald Charles and other geneticists, including L. H. Snyder, George Beadle, and C. H. Danforth. Lewis Weed represented the NRC, as did several staff members, including Hayden C. Nicholson, Philip S. Owen, John C. Ransmeier, and John J. Lentz, Jr.[31]

The meeting was held at Neel's insistence to "review and appraise" his plans for the genetics project. It provided a forum for him to obtain feedback from more experienced senior scientists. For seven hours on 24 June 1947, this group digested and discussed Neel's various reports on his groundwork in Japan.[32] The initial proposal was roughly an extension and improvement of a plan already in effect in Hiroshima, devised by Matsubayashi Ikuzo, a Japanese physician interested in tracking genetic effects of the bombs.[33]

Matsubayashi, who worked for the Section of Public Health of the Hiroshima city government, had begun a genetics study in the fall of 1945 under the National Research Council of Japan's Special Committee for the Investigation of Atomic Bomb Effects. He was tracking pregnancies in Hiroshima through a special food rationing registration program made necessary by postwar shortages in Japan. Pregnant women qualified for additional food rations beginning in the fifth lunar month of their pregnancies. Matsubayashi was using this registration program to contact all pregnant women before their babies were born. Neel had heard that "certain 'upper class' or wealthy women" neglected this registration, because they did "not need the added allowance." He also suspected that some women might be registering twice, for increased rations, or registering when they were not in fact pregnant (though the city required a doctor's certification of pregnancy).[34]

Despite the risk of biased sampling, the pregnancy registration program

31. Neel's draft report prepared for this conference, "An Evaluation of the Feasibility of Studies on the Question of the Genetic Effects of the Atomic Bombing," is in his personal papers, 5/ B16, JVN. The conference report, "Concerning the Study of the Genetic Effects of the Atomic Bombs in Hiroshima and Nagasaki," 30 June 1947, is in drawer 19, Genetics—1947 #1, NAS; "Minutes of the Conference on Genetics," 24 June 1947, 5/BB, JVN. Also, see National Research Council, Committee on Atomic Casualties, Genetics Conference, "Genetic Effects of the Atomic Bombs in Hiroshima and Nagasaki" (1947).

32. See "Summary of Proceedings, Conference on Genetics," 24 June 1947, 5/BB, JVN.

33. Matsubayashi was also responsible, with statistician Masato Kano, for carrying out a population survey in Hiroshima beginning shortly after the bombing. Tsuzuki Masao had suggested the survey as a means to obtain a record of all survivors, as well as descriptions of where they were at the time of the blast and of their injuries. With the help of volunteers who worked each city block, Matsubayashi and Masato obtained records on 145,000 of the 200,000 persons in Hiroshima in August 1945. See Atomic Bomb Casualty Commission, "Report No. 2, for the period 2–7 December 1946," 6/A, JVN.

34. Ibid., 2.

did make possible a prospective study of pregnancies and births. The ABCC staff could interview women about their radiation exposure before the women knew how their pregnancy would end. This might enhance the reliability of the exposure histories. The plasticity of human memory is a well-recognized phenomenon in epidemiology, and prospective studies, in which information is obtained from individuals before the relevant outcome is known, are considered more powerful tests of a hypothesis. In a study such as this women interviewed after an undesirable pregnancy outcome—a neonatal death or the birth of an abnormal child, for example—might be expected to search their memories for some explanation, perhaps radiation exposure, that could make sense of a personal tragedy. If they were interviewed in the early months of the pregnancy, however, their account of radiation exposure could at least be presumed to be uninfluenced by the particular outcome of their pregnancy. Other influences—the desire to conceal one's status as a bomb survivor because of discrimination, for example—could not be ruled out, of course. Some participants did revise their accounts of what had happened to them when the bombs were dropped, and some of these revisions came after their pregnancies ended. The ABCC generally accepted these revisions. Yet Neel later assessed the existence of the rationing system, and the access it provided to women in the early stages of pregnancy, as the single most important factor in the feasibility of the genetics studies in Hiroshima and Nagasaki.[35] Certainly it was one of the few elements in the social and economic chaos of immediate postwar Japan that promised to facilitate the proposed scientific study.

Matsubayashi had prepared a questionnaire to be partially filled out during a first interview at City Hall and completed by midwives upon termination of the pregnancy. After the obvious questions—where were you when the bomb exploded? what were your injuries?—it included a section in which the outcome of the pregnancy could be described by the midwife or, more rarely, the delivering physician. No similar program was planned for Nagasaki.[36]

Neel felt there were some defects in the Japanese program, but "so far as it goes it is soundly conceived." However, the Japanese had failed to make provision for a suitable control population. Matsubayashi planned to use Japanese vital statistics from the past few years to compare sex ratio and rates of stillbirth, malformation, and miscarriage. Neel was skeptical of the value of such statistics, suspecting that, while the simple birth and death registrations were probably relatively complete, the "reporting of abortions and miscar-

35. James V. Neel, interview with author, 4 February 1988, Ann Arbor, Michigan.

36. See James V. Neel, "An Evaluation of the Feasibility of Studies on the Question of the Genetic Effects of the Atomic Bombing," draft report, second revision, dated mid-May 1947, 5/ B16, JVN.

riages has been somewhat less accurate."[37] This would tend to exaggerate any genetic effects of radiation among the survivors. He also believed that infanticide might be fairly common in Japan in cases of certain congenital anomalies. But with only hearsay evidence he could not be sure what effect infanticide might have on the accuracy of official records or on the genetics study.[38]

The aspects of the Japanese study that Neel incorporated into his own proposal included the use of a two-part questionnaire, somewhat more detailed than Matsubayashi's, and the use of the midwife collection system. Both Matsubayashi's study and the later ABCC study facilitated midwife cooperation by paying the midwives for each birth reported.[39] Matsubayashi paid only one yen in 1948, while the ABCC paid ten, with fees increasing through 1954.[40]

Neel's plans in May 1947 were far-reaching and ambitious, going well beyond the Japanese study already in effect. He suggested that all abnormal terminations be followed up with "detailed family studies"; that a system of dual registration, to allow cross-checking of birth reports, be set up for midwives, physicians, and families; and that the offspring of irradiated persons be periodically reexamined to detect and record genetic effects not apparent at birth.[41] However, because of practical limitations, family analyses were not done, dual registration was never established, and "repeated examination" was limited to one follow-up visit, at nine months of age, for about 10 to 30 percent of the sample.

Neel's plan was well thought out and organized, and when he presented it to the distinguished scientists called together to review it he must have been pleased with the response. While the conference members expressed concerns about the problems inherent in the situation, they supported his plans and endorsed his scientific approach.[42]

Muller, the dean of radiation genetics, was sympathetic to Neel's plan but wondered whether the study should be done at all. His objections were similar

37. Ibid., 4.

38. James V. Neel, "Concerning the Study of the Genetic Effects of the Atomic Bombs in Hiroshima and Nagasaki," report submitted to the subcommittee on genetics of the National Research Council, 10 June 1947, JVN.

39. Neel describes Matsubayashi's program in his essay prepared for the conference, "An Evaluation of the Feasibility of Studies on the Question of the Genetic Effects of the Atomic Bombing," draft report, second revision, dated mid-May 1947, 5/B16, JVN.

40. Conference of the Genetics Group, "Memorandum for the Record," 5 June 1948, Aga Machi, drawer 19, Genetics—1948 #2, NAS.

41. James V. Neel, "An Evaluation of the Feasibility of Studies on the Question of the Genetic Effects of the Atomic Bombing," draft report, second revision, dated mid-May 1947, 5/B16, JVN.

42. Ibid., 29.

to those raised earlier in response to Paul Henshaw's proposed survey of radiologists. Muller feared it was a "dangerous situation from the standpoint of the general and scientific public, who would be prone to assume that if no effect could be demonstrated, there must not have been one." It was "certain" on theoretical grounds that some genetic damage must have resulted from the bombings, Muller said, but he suspected that very little of this effect could be demonstrated under "this or any other plan." He expected the rate of undetectable mutations to be higher than the rate of detectable ones. In his own work with *Drosophila,* he had estimated that, of fifty sex-linked mutations, thirty-nine would be "invisible," ten would be lethal, and one or two would be visible recessives.[43]

Muller and Neel tried to estimate how many new mutations would have been produced at Hiroshima and Nagasaki if the sensitivity of human genes to radiation mutagenesis were identical to that of the mouse, basing their calculations on Charles's estimates for mouse mutation rates. They figured that about twelve thousand children would be born in the next ten years to parents one or both of whom had been within two kilometers of the hypocenters. Neel estimated that the upper dose of radiation received by the survivors would be about 500 rads. This was the dose at which approximately half of those exposed would die. Assuming a mean dose of 300 rads in the parents, they calculated the total exposure as 300 rads times 12,000 offspring, or 3.6 million rads. With a dominant mutation rate identical to that of the mouse, this would correspond to thirty-six to seventy-two new mutations resulting in gross abnormalities. These would be added to the one hundred twenty abnormalities expected under normal conditions in this many births. As Neel observed, many extraneous factors could obscure such an increase.[44]

Recessive, or "invisible," mutations, which might be expressed in later generations, would be virtually inaccessible in the children of the survivors. With this in mind, Muller suggested that a well-planned mammalian study paralleling the studies of human populations in Japan might provide some insight into the frequency of recessive mutations. Such a study could focus on the same biological events being tracked in Japan. The parallel study might shed light on the proportion of mutations undetectable in human populations.[45] This suggestion provoked an enthusiastic discussion of the need for animal studies since, as Beadle put it at a meeting two days later, "from a purely scien-

43. "Summary of Proceedings, Conference on Genetics," 24 June 1947, 5/BB, JVN.

44. Ibid., 30.

45. This suggestion was not followed. In fact, no mammal studies of analogous genetic endpoints has yet been carried out in parallel with the early genetics studies in Japan.

tific standpoint" more could be learned from monkeys than from the study of the Japanese survivors.[46]

Alone among the participants in the genetics conference, Don Charles suggested an entirely different approach. Instead of tracking all births in Hiroshima, Nagasaki, and a control city, he said Neel should select three types of representative matings, and compare the chosen parameters among them. These were matings between two irradiated parents, matings between an irradiated and a nonirradiated parent, and matings between two nonirradiated parents. The control city, Kure, would not then be necessary. Charles also suggested some sort of incentive for long-term Japanese cooperation was necessary and thought that incentive might be medical treatment of the participants. And he felt that the ABCC should turn the project over to the Japanese as rapidly as possible. Though the committee did not choose to endorse Charles's approach, his comments were remarkably prescient. Kure was abandoned in 1950, no longer deemed necessary due to the presence of internal controls. Treatment became a major issue and stumbling block for the ABCC. And the role of the Japanese—their relative status and power in the ABCC—was an important issue not resolved until the Atomic Bomb Casualty Commission became the Radiation Effects Research Foundation in 1975.

Neel's plan for a major epidemiological survey of all births to exposed and control populations was supported by the committee at large, despite gloomy predictions about the possible scientific results. There was "some discussion as to the extent to which our thinking was being influenced by 'non-scientific' considerations."[47] These included the "irresponsible conjectures concerning the genetic effects of the bombs" which were "widely circulated" at the time.[48] Such considerations must have been sufficient to overcome the low expectations for the project; the group endorsed the plan for a genetics study and agreed to announce publicly that significant results were not expected.

Their statement in *Science* in October 1947 said that although genetic effects would probably not be demonstrable, this "unique possibility for demonstrating genetic effects caused by atomic radiation should not be lost." They assumed that the need for an American study, rather than one involving scientists from many countries or a strictly Japanese study, was obvious—at least, the issue was not explicitly addressed. The statement merely noted that Japanese studies were already in place, but were "under great difficulties" when

46. Committee on Atomic Casualties, "Minutes," 26 June 1947, NAS.

47. Ibid., 30.

48. The phrasing here comes from Neel's 1985 account in "Delayed Biomedical Effects of the Bombs" (Neel, Beebe, and Miller 1985).

the ABCC study team arrived, and that a "good deal of misinformation" about genetic effects was circulating in Japan. Although genetic effects had presumably occurred, the article stated, "the conference wishes to make it clear that it cannot guarantee significant results from this or any other study on the Japanese material" (Genetics Conference, Committee on Atomic Casualties 1947).

The Committee on Atomic Casualties, the board newly appointed by the National Academy of Sciences to oversee the work of the ABCC, met two days later to add its endorsement. One committee member, Raymond Zirkle, said the study would be "worthwhile from the standpoint of the morale of people who are working at tasks involving radiation hazards." Presumably their morale would not be improved by results suggesting dramatic genetic effects could be expected in their children; Zirkle was apparently assuming that the final results would be negative. The study would "remove much uneasiness" and prove "comforting" to the public who would know that competent scientists were investigating the possible genetic effects of the bomb, he added.[49]

Prior to these two meetings, Neel had occasional doubts about whether the study should be done at all. He wondered if something should be published "as a sort of explanation and justification" for not going through with the genetics project. Neel recognized that there might be legitimate reasons for such a decision. By late June, however, the consensus was that the project had to be done.[50] Bolstered by the endorsements of so many respected scientists, Neel wrote confidently back to Japan that the genetics project would soon begin.[51] He was not optimistic about detecting mutations, but he felt the study could provide "considerable basic data on what might be called 'conception outcome in Japan'" and "in the end this data may be more important than what we find out, or do not find out, about the bomb effects."[52]

CONCLUSION

When it began, the genetics project was expected to fail. Its planners believed that the study would be unable to demonstrate significant genetic effects in the survivors and their children. At the same time, they believed that genetic effects had occurred. Mutations could be present, but undetectable, for several

---

49. Committee on Atomic Casualties, "Minutes," 26 June 1947, NAS.

50. James V. Neel to Austin M. Brues and Paul Henshaw, 13 March 1947, 6/J, JVN.

51. James V. Neel to Harry Johnson, 29 June 1947, 6/J, JVN. Johnson was a colonel with the Public Health and Welfare Section of the Occupation.

52. James V. Neel to Harry Johnson, 7 August 1947, 6/J, JVN.

reasons. If a mutation caused a sperm or an egg to be nonviable, or nonfertilizable, or a fertilized egg to be spontaneously aborted, or if the mutated gene were repaired, a "mutation" (change in the hereditary material) would have occurred but it would have no impact on living offspring *of any generation.* Mutations could also be undetectable (but *present*) if they were recessive, affecting the hereditary material but not expressed in the first generation born after the bombings. And mutations could be undetectable for epidemiological reasons: the large number of births involved, the relatively small number of expected abnormalities, and the general uncertainties inherent in diagnosing abnormalities in infants, detecting spontaneous abortion, and acquiring radiation histories.

The *Science* essay's closing statement therefore dealt with at least three levels at which "mutation" could be defined. At the molecular level a mutation was any change in the hereditary material. At the organismal level, it was an abnormality of structure or function detectable in a child. At the population level, a mutation was a statistical construct: a significant difference in the rate of abnormalities in the children of parents exposed to the bomb's radiation, and those who had not been exposed. The ABCC genetics study was shaped by uncertainties at all three levels.

The study of human inheritance in general was problematic in the 1940s for at least two reasons. First, the methods of genetics until the 1950s depended primarily on selective breeding to infer genetic structure, and human populations cannot be selectively bred. Pedigree studies of blood types, Huntington's Disease, Tay-Sachs Disease, and Down's syndrome had been reasonably successful, but these were meager findings when compared to the productivity of work on *Drosophila,* which by 1940 had yielded gene maps and giant salivary chromosomes. Human inheritance did not lend itself to carefully controlled or "rigorous" scientific studies, and while its findings were in some ways more socially valuable than those from investigations of *Drosophila,* its potential for gaining insight into the fundamental nature of heredity was limited.[53]

Second, many published works that did deal with human heredity were infused with social agendas their promoters made no attempt to conceal. From Francis Galton in the late nineteenth century to H. J. Muller in his famous 1950 essay "Our Load of Mutations," human heredity was often described in terms of its social implications (see Muller 1948; Muller 1950a; also Carlson 1981; Paul 1987). An understanding of human inheritance was supposed to be useful in controlling human reproduction for the good of future generations, or for the improvement of the course of human evolution. This was the goal, more or

53. Curt Stern provides valuable summaries of the changing states of human genetics in his first and second editions of *Principles of Human Genetics* (1950, 1960).

less, of the various eugenics movements that flourished around the world in the early twentieth century. These movements resulted in laws ordering sterilization of the "unfit," restriction of immigration into the United States of ethnic groups deemed inferior, and, in Germany, medical termination of "lives not worth living," that is, the murder of mental patients, deformed children, and others by nurses and physicians acting in accordance with Nazi law (Kevles 1986; Proctor 1988). With rising public awareness of Hitler's racial policies, eugenics had become, by 1940, a socially unacceptable field.

Human genetics was not technically eugenics—though many who studied human heredity and that of other organisms were eugenicists. But geneticists of all kinds were influenced by the negative popular image of eugenics, and students interested in human heredity were often encouraged to turn their attention elsewhere (Kevles 1986). Human genetics was therefore, in 1946, both methodologically and socially suspect. Human genetics was not widely taught in American medical schools until the early 1960s and it promised few professional rewards to its practitioners. It was not particularly useful in any "practical" way, as a guide to genetic counseling, for example, since it could offer prospective parents only statistical probabilities for a few rare conditions.

Yet despite these problems the postwar period saw rapid growth and change in the status of human genetics. The American Society of Human Genetics was founded (1948), the *American Journal of Human Genetics* began publication (1948), and the first human genetics program in an American medical school was established (University of Michigan, 1952).[54] In 1948, only a dozen or so American geneticists identified themselves as human geneticists, and not all of these worked on human populations. By 1960, hundreds of students of genetics had turned their attention and scientific resources to solving problems of human heredity.[55] Many who started out in corn or fly genetics shifted in the late 1950s, as the changing scientific and professional climate made human genetics an increasingly attractive field.

In this climate of rapid change, the genetics project at Hiroshima and Nagasaki was an important demonstration of the practical value of more scientific approaches to human heredity. It was the largest epidemiological project of its kind up to that time—a classic example of "big science" in biology—and it dealt with a subject of pressing national and international concern. The genet-

---

54. Madge Thurlow Macklin, at Ohio State University, championed the introduction of medical genetics into the standard medical school curriculum throughout the 1940s and 1950s. She argued that physicians needed to understand genetics in order to diagnose and treat (Macklin 1941, 1948).

55. For the early years of the American Society of Human Genetics, the scrapbook of Herluf Strandskov is useful (Ms. Coll. 49, APS).

ics project was the largest and most interactive of the ABCC's programs. It involved contacting pregnant women, learning their radiation exposure history, and examining their children at birth and, in some cases, at nine months. The system established to accomplish this was complicated and expensive and depended heavily on the engagement of Japanese staff, midwives, and subjects of research.

 **TWO**

*Managing
the
ABCC*

CHAPTER FIVE

Midwives and Mothers

T he relations of American Occupationaires to the Japanese public have sometimes been romanticized in both personal memoirs and American fiction. Elizabeth Grey Vining's 1952 *Windows for the Crown Prince,* an account of her 1946 to 1950 term as tutor to Crown Prince Akihito (now emperor of Japan), was a best-seller, as was James Michener's *Sayonara* two years later.[1] American reminiscences have presented the Occupation as a time of American goodwill and enchantment with Japanese art and culture.[2] ABCC

1. Sheila K. Johnson explores this phenomenon in *The Japanese through American Eyes* (1988, 55–72).

2. See particularly Grant K. Goodman, ed., *The American Occupation of Japan: A Retrospective View* (1968); also, Harry Emerson Wildes, *Typhoon in Tokyo: The Occupation and Its Aftermath* (1954). The most influential anthropological account is Ruth Benedict, *The Chrysanthemum*

geneticist William J. Schull has recently provided his own account, *Song Among the Ruins*, which celebrates both America's good intentions and the appeal of Japanese culture. Schull's story, however, also describes a fair amount of misunderstanding, confusion, and tension between Americans and Japanese in Occupied Japan.[3]

The historical assessment of the Occupation—its long-term impact on Japanese culture and society—is a subject of continuing debate among scholars in both the United States and Japan (Ward and Sakamoto 1987; Gluck 1983). The "heroic narratives" of Occupation scholarship have often focused on the actions of General Douglas MacArthur or those of his chief lieutenants (Willoughby and Chamberlain 1954; Whitney 1956).[4] In the past decade, however, MacArthur as the hero who transformed Japanese society by the force of his personality has given way to MacArthur as an important, but by no means all-powerful, player in a complex bureaucracy. Some scholars have begun exploring the degree to which SCAP policies were formulated "from the ground up," the product of individual decisions by lower-echelon American bureaucrats or even Japanese advisers. Policies that once seemed to have derived from MacArthur alone are now beginning to be tracked to other individuals with their own agendas and goals (Ward and Sakamoto 1987).

Susan J. Pharr's study of women's rights reform during the Occupation, for example, traces the origins of Japan's constitutional amendment granting women political rights to a twenty-two-year-old woman who served on a three-member committee that drafted a bill of rights in 1946. Beate Sirota had lived in Japan from the time she was five until she was fifteen, was fluent in Japanese, and, because her father headed the piano department of the Japanese Imperial Academy of Music in Tokyo, knew many Japanese women artists and perform-

---

and the Sword: Patterns of Japanese Culture (1946). For a perceptive early view of the Occupation, see Kawai Kazuo, *Japan's American Interlude* (1960). The author was Japanese-born and a scholar of Japanese politics in the United States both before and after the war. He spent the war years in Japan and during the Occupation was editor in chief of the *Nippon Times* (today the *Japan Times,* Japan's leading English-language newspaper) from 1945 to 1949. Kawai suggests that the relative success of the Occupation was grounded in the preexisting democratic sympathies of the Japanese. For a relatively recent celebration of the Occupation as a vindication of American ideals, see John Curtis Perry, *Beneath the Eagle's Wings* (1980). Perry, however, also links the American experience in Japan with the hubris that led to American involvement in Vietnam. An excellent review of the changing historiography of the Occupation is Carol Gluck's article, "Entangling Illusions: Japanese and American Views of the Occupation" (1983).

3. Schull clearly expresses his dissatisfaction with some aspects of Japanese culture, however (Schull 1990).

4. Gluck refers to these accounts as "heroic narratives," borrowing the term from Butterfield (Gluck 1983).

ers. She therefore had particular insight into the needs of Japanese women. Sirota personally drafted what became Article 24 of the Japanese constitution, establishing the "essential equality of the sexes" in matters of property rights, inheritance, divorce, choice of spouse, and other matters pertaining to marriage and family. While MacArthur supported this provision and later emphasized his personal commitment to women's rights, this support came after the fact. None of MacArthur's directives to those drafting the constitution mentioned women or women's rights, and Sirota's actions seem to have been the product of her own interests and convictions. These interests had significant consequences (Pharr 1987).

As such recent scholarship suggests, the Occupation's impact might profitably be studied through an analysis of the day-to-day interactions of individual Americans and Japanese. For the residents of Hiroshima and Nagasaki, these interactions were typically with the American staff of the Atomic Bomb Casualty Commission. Many survivors of the atomic bombings had their most important and sustained contact with Americans through the ABCC—particularly its genetics project.[5] Most of the day-to-day work of this project was carried out by Japanese midwives, pediatricians, survivors and controls, and clerks. American scientists made policy and administrative decisions, American physicians handled questionable diagnoses, American geneticists analyzed the genetics data—in Ann Arbor, Michigan—but the system of data collection was almost entirely Japanese.

THE GENETICS SURVEY

To track births in Hiroshima and Nagasaki, the ABCC took advantage of a postwar rationing system in which pregnant women qualified for additional food starting in their fifth month. When a pregnant woman came to register for these rations at City Hall, the city clerk asked her also to stop by the ABCC office to register for the genetics program. Municipal officials in Hiroshima, Nagasaki, and Kure provided the ABCC with office space. In this way the ABCC was able to register more than 90 percent of all pregnancies in the three cities.[6]

5. Many staff sections of SCAP had relatively few direct dealings with the Japanese public. They worked rather through the Japanese ministries to which they were assigned (Kawai 1960, 19).

6. A study of the unregistered pregnancies in October 1952 suggested that those not registering had either arrived in the city just prior to giving birth, had registered at the city hall for food rations but failed to register with the ABCC, or had failed to register for food rations because they forgot, were sick, did not know about the rations, and so on. Many of these unregistered pregnancies came to the attention of the ABCC through midwives, who reported the terminations, or

When pregnant women came to this ABCC office (and the majority did) the clerk gave them a pamphlet, "To All Prospective Mothers," which outlined the genetics project plans. The Japanese text, according to the translation in Neel and Schull's 1956 book, opened with the presumption that the pregnant woman was "familiar with the research project of the ABCC." If the woman was not familiar with the ABCC and its research, she would not be enlightened by the text itself, which did not state that the study was an effort to track the genetic effects of radiation from the atomic bombs. It reassured the registrant that she "need not be worried about the questionnaire," for it contained only "questions concerning your expected baby, name of parents, birthdates, date of marriage and history of exposure to the atomic bomb." The questionnaire, however, also included personal questions about reproductive history, including numbers of therapeutic abortions, stillbirths, and miscarriages. The pamphlet went on to explain that an ABCC doctor would come by to see the baby after the birth, or to talk with the mother in case of stillbirth or neonatal death. The text stressed both the "modern x-ray and other medical equipment" available at the ABCC and the advantages to the mother: "By this examination you will be able to know your baby's true physical condition and at the same time you will be making an important contribution to medical science." In addition, "all services are performed free of charge."[7]

After she read this pamphlet the pregnant woman spoke with the ABCC genetics clerk at City Hall. The clerk, a Japanese employee of the ABCC, took the woman's radiation history, asking where she had been at the time of the bombings. Many residents of Hiroshima and Nagasaki in the years after the war were new to the city and had not been exposed to the bombs—these were eventually designated as controls for the genetics survey—but if a woman reported that she was exposed to the bomb, the clerk asked her to specify precisely her location. What street address had she been at? Was she indoors? What sort of building was she in? He also asked if she had experienced any symptoms of radiation sickness, such as hair loss, subcutaneous bleeding, gingivitis, or bloody diarrhea, or sustained any burns or other physical injuries. The details of this radiation history were used by the ABCC to classify survivors in various categories of radiation exposure (Neel and Schull 1956a, 6–7).

This interview took place before the pregnancy ended and before any un-

---

through physicians. See *The Effect of Exposure to Atomic Bombs on Pregnancy Termination in Hiroshima and Nagasaki* (Neel and Schull 1956a, 7).

7. The pamphlet is reprinted in Neel and Schull (1956a, 8). A slightly different (more awkward) translation is "Literal Translation of Japanese Pamphlet Distributed at Time of Pregnancy Registration," 5/QB, JVN.

desirable pregnancy outcome could lead the woman to "remember" proximity to the bomb. The ABCC staff therefore accepted the radiation histories obtained in these interviews as accurate. However, many of the questions were open to broad interpretation. Symptoms of acute radiation exposure, for example, can vary considerably. Hair loss, a key sign of radiation sickness, can range from a few handfuls to total baldness. Strictly speaking there was no way of distinguishing these two extremes based on the yes-or-no answers obtained in this original interview. The ABCC clerk asked if there had been hair loss, but not how much. The same was true of other symptoms, including gingivitis, subcutaneous bleeding, and bloody diarrhea.

In addition, the clerk asked the women to provide similarly detailed information about their husbands' history of radiation exposure. This may have been a major weakness in the data collection plan. The ABCC asked each woman—whether she was a survivor or not—to report her husband's exact location at the moment of the explosion of the bomb, his degree of shielding, and his symptoms in the weeks after the bombings. Virtually all information about paternal exposure in the ABCC morphological genetics study came from the wives: the genetics clerks did not interview male survivors. All reports on paternal irradiation therefore had the disadvantage of being secondhand.

Answers about reproductive history at least had the advantage of coming from the person involved. A woman might be expected to remember her own pregnancies, abortions, miscarriages, or stillbirths. The personal nature of the questions, however, might have discouraged direct answers. The ABCC wanted to know if the husband and wife were related to each other, how long they had lived together, the total number of pregnancies a woman had experienced, and whether she had experienced any abortions, spontaneous or induced. A woman might have many possible motives for not answering such questions directly.

At the end of the interview, the clerk retained a copy of the partially completed questionnaire and gave a copy to the pregnant woman. Thirty-seven of the fifty questions were answered in this initial interview at City Hall. The remaining thirteen questions concerned the outcome of the pregnancy. The midwife, or occasionally a physician, was to fill these in when the pregnancy ended. When a woman registered her pregnancy, she was asked which midwife would be delivering her child. The ABCC kept cards for each midwife practicing in the city, and the woman's name and expected date of delivery would be entered on her midwife's card.[8]

8. George A. Hardie, "Report of a Visit to ABCC in Japan, June–July 1950 [Copy No. 3]," NAS, 132.

RELATIONSHIPS WITH THE MIDWIVES

The Japanese midwives were the contacts that linked the ABCC to the subjects of the genetics study. Most births in immediate postwar Japan (various estimates place this number at 85 to 96 percent) were attended by a midwife rather than a physician, and the licensed, trained midwives were well-known community members with strong ties to the families they served. They provided the ABCC with access to women bearing children in Hiroshima and Nagasaki. The cooperation of the midwives was critical to the genetics study, and the ABCC staff invested much time and energy in improving communication with this large group of women. The staff also tried to improve the accuracy of reporting by educating the midwives to recognize abnormalities and encouraging immediate reporting of unusual births. The ABCC paid the midwives for each birth reported, with a bonus for quick reporting of those that appeared to be abnormal.[9]

The Americans' relationship to the midwives was complicated by both culture and gender. The two groups had different notions of acceptable behavior. Issues that concerned the midwives deeply, such as the selection of a crematorium, were of relatively little interest to the American scientists and physicians. And certain American priorities, such as the rapid reporting of stillbirths, were difficult to convey to the midwives, who had their own sometimes compelling reasons for delaying such reports.[10]

Frequent, organized contact played an important role in resolving some of these differences. In the summer of 1948 alone ABCC staff members met eighteen times with groups of midwives in the Hiroshima-Kure area. From six to thirty midwives attended each meeting. The ABCC staff was usually represented by a female Japanese secretary; two male scientists from the Japanese NIH, with which the ABCC was affiliated; and at least four ABCC physicians, usually three Japanese and one American, all male.[11] The importance attached

9. Midwives were also used to collect information in Germany when Nazi racial policy encouraged physicians to kill infants with significant abnormalities. The midwife as gatekeeper deserves further study. See Robert Proctor, *Racial Hygiene: Medicine under the Nazis* (1988).

10. One of my readers has suggested that I need to deal here with the attitudes of Japanese physicians and scientists employed by the ABCC. But it is difficult for me to assess the viewpoint of the ABCC's Japanese physicians and scientists on these questions, since most of the records I have been able to examine that deal directly with the midwives were produced by Americans (Neel, Tessmer, McDonald, Ray Anderson, and others). While the words and actions of the midwives are often recorded in these accounts—these are my clues to their perceptions and expectations—the responses (if any) of Japanese scientific staff are generally not mentioned.

11. Ray Anderson, "Memorandum for the Record," "Meetings with Midwives of Hiroshima and Kure during June–July 1948," 15 July 1948, JVN.

to these meetings is suggested by ABCC director Carl Tessmer's frequent attendance. Tessmer had many other duties and obligations at this time, including negotiations with municipal officials in Hiroshima and Kure, personnel matters, meetings in Nagasaki with representatives of the medical community, and such mundane matters as tracking down furniture and office equipment.[12] Yet he took the time to sit through repetitive meetings with various groups of Japanese midwives, partly because his presence demonstrated the importance the ABCC attached to midwife cooperation.

In general the meetings with midwives consumed a great deal of professional staff time, often on activities that did not require professional training. Physician Ray Anderson and geneticist Masuo Kodani, for example, often chauffeured midwives to meetings at the Gai-Sen-Kan in Hiroshima (the intermediate offices of the ABCC prior to the completion of the Hijiyama facilities) in an ABCC station wagon. The staff found that both providing transportation and serving a light lunch "contributed greatly to the success or failure of individual meetings," and this justified the efforts of the scientific staff.[13]

In the spring of 1947, Neel spoke at a meeting of the Hiroshima Midwives Association and the Prefectural Midwives Society and emerged confident of the midwives' interest in the project. He decided to have special membership cards printed on heavy, glossy paper, and he planned to conduct a short course on the classification and terminology of deformities for the midwives, in the hope that such training would enhance the accuracy of their birth reports.[14]

Japanese midwives in 1946 were relatively well educated, having completed eight years of schooling to qualify for licensure. They were not trained as physicians but had spent at least two years in medical training, either one year in a medical school and one as apprentice to a more experienced midwife, or two years at a medical school. Candidates also had to pass a prefectural examination (Neel and Schull 1956a, 5). While the midwives could not be expected to recognize and classify all birth anomalies they did at least have some knowledge of medical terminology and, perhaps more importantly, thought of themselves as professionals with high standards to maintain. The midwives' pride in their work and professionalism provided one motive for their cooperation with the ABCC. Many apparently felt that they were participating in an important scientific project, and some said they had gained professional prestige by their association with the American study. The Nagasaki

12. Particularly Tessmer's letter detailing his activities, Carl Tessmer to James V. Neel, 7 July 1948, drawer 19, NAS.

13. Ray Anderson, "Memorandum for the Record," "Meetings with Midwives of Hiroshima and Kure during June–July 1948," 15 July 1948, JVN.

14. James V. Neel to Richard Brewer, 18 November 1947, JVN.

group felt they were "raised in comparison with other midwife associations in Japan" as a result of their work with the ABCC.[15]

The midwives held a position of status "similar to that of the old family physician in this country." Their relationship to the family could last many years and be extremely close. Many women gave New Year's presents to their midwives and invited them to the various parties and celebrations that accompanied a pregnancy and birth in 1940s Japan. It was customary for the midwives to send no bill for their services. Rather, the patient was expected to give the midwife a cash "present," determining the appropriate amount by asking friends and relatives. In 1950, the customary rate for a delivery was two thousand yen.[16]

The midwives were not well informed about the ABCC genetics program when the meetings with Neel began. Most listened attentively and some raised lively questions, though there were a "few at each meeting" who dozed off.[17] Overall, the image conveyed of the midwives' participation in these early meetings was of a group whose members politely consented to attend because they were asked, but who were not yet particularly committed to participating in the scientific study. The ABCC needed a much more serious commitment from the midwives if the genetics study was to succeed. One way to encourage their cooperation was to pay them for every birth report. Matsubayashi had paid midwives one yen for every report; the ABCC proposed to pay them five or ten yen per questionnaire. The midwives did not object. They asked only that they be allowed to return the completed forms (reporting pregnancy outcomes) on the fifth of each month, rather than on a weekly or daily basis.[18]

By 1951, the ABCC's schedule of fees included bonuses for quick reporting of abnormal births. The ABCC paid fifty yen for a report of a normal termination, and up to two hundred yen for a report of a neonatal death within twelve hours of the death.[19] A reasonable (low-range) monthly salary for a newly trained physician in 1950 in Hiroshima and Nagasaki was about five thousand yen. At the other end of the scale, a Japanese driver-handyman at the Tokyo ABCC office earned approximately nine hundred yen—or about thirty-

15. D. J. McDonald to Newton Morton, Grant Taylor, and James V. Neel, 2 December 1953, drawer 19, Report on a Midwives Meeting 21 November 1953, NAS.

16. George A. Hardie, "Report of a Visit to ABCC in Japan, June–July 1950 [Copy No. 3]," NAS, 132.

17. Ray V. Anderson, "Memorandum for the Record," "Meetings with Midwives of Hiroshima and Kure during June–July 1948," JVN.

18. Ray Anderson, "Memorandum for the Record," 26 January 1948, in First Meeting of the Hiroshima-Kure Advisory Committee, JVN.

19. "Meeting, Hiroshima Midwives Association," 22 September 1951, 5/BC, JVN.

four dollars—per month in 1950.[20] The standard payment by the family for a midwife delivery was two thousand yen in 1950, so the monthly income of a midwife who delivered two infants per month would be four thousand yen. A midwife reporting one normal birth and one abnormal birth in a single month would enhance her monthly income by three hundred yen, or eight percent. Successful midwives could expect to deliver more than two infants: Yamamoto Setsuko of the Hiroshima Midwives Association estimated that midwives averaged ten deliveries a month in 1949.[21] Even ten normal reports would increase a midwife's income by five hundred yen, and if any of the infants were abnormal the payment would be much higher. The ABCC put a cap of fifteen hundred yen on the amount payable to a single midwife in a single month; the necessity for such a limit suggests that at least some midwives were coming close to or exceeding this amount in fees.[22]

Some of this income may even have been tax-free. In September 1951 a "frightful thing" happened in Nagasaki. A "tax collector got admission to the midwives' payment ledger . . . and started to use the information he obtained to bludgeon the midwives into payment of taxes." ABCC geneticist Duncan McDonald supposed that "a few of them had been concealing part of their income." Masuo Kodani, a Japanese-American physician in the Nagasaki ABCC office, "obtained an apology from the Tax Office and the chap who obtained and used the information was made to deny it to those against whom he had used it. Not a very nice situation, but we have made sure it won't happen again."[23]

In any case, these small fees could cumulatively make a real financial difference to the midwives. The payments certainly played a role in ensuring high levels of cooperation. At the same time, cooperation with the ABCC raised some inherent conflicts for the midwives. The birth of an abnormal infant was a source of family shame in Japan. It might even adversely affect the marriageability of other family members, precluding a desirable match. There was, therefore, considerable motivation to conceal the birth of an abnormal

20. George A. Hardie, "Report of a Visit to ABCC in Japan, June–July 1950 [Copy No. 3]," NAS, 38.

21. Yamamoto Setsuko, head of the Hiroshima midwives organization, estimated this delivery rate for an average midwife in a meeting with W. J. Schull in 1949. "Memorandum for the Record," 15 December 1949, JVN.

22. Payments to midwives came through the Japanese NIH. In February 1950 they totaled 13,610 yen in Hiroshima, 9,620 yen in Kure, and 14,980 yen in Nagasaki. The limit payable to any individual midwife is mentioned in George A. Hardie, "Report of a Visit to ABCC in Japan, June–July 1950 [Copy No. 3]," NAS, 133.

23. Duncan McDonald to James V. Neel, 7 November 1951, 6/T, JVN.

child. The midwives had a "strong allegiance to the family," Neel noted after his first visit to Japan, and might not report abnormalities because of the "great deal of prejudice against having the occurrence of an abnormal birth made known." [24] While the ABCC made every effort to preserve the anonymity of study participants in their formal records (through numerical identification of each subject), this system did not completely overcome the reluctance of families to tell ABCC representatives—physicians, clerks, and others—about abnormal births or early spontaneous abortions.

A preliminary study in the summer of 1948 suggested that the midwives were failing to report many malformations, though the report did not speculate on their motives. "There has been a feeling all along that the midwives and doctors would not be reporting all abnormal births, and this is confirmed by the data," the report stated. Of 258 "normal" births in Hiroshima, 12 (5 percent) proved to have abnormal features. In Kure the corresponding figure was 12 of 144 (8.3 percent).[25] By 1949 the ABCC staff had confirmed this early finding. An ABCC staff member complained of a "distinct general tendency to conceal malformations." [26] While many of these abnormalities were picked up in ABCC home visits, or in examinations of a 10 to 30 percent sample at the age of nine months, the discrepancy did suggest a fundamental problem in the midwife data collection system. In the case of stillbirths, for example, an abnormality not recorded by the midwife would be missed, particularly if the ABCC was not notified of the stillbirth until after the body had been cremated.

A midwife reporting a major malformation to the ABCC knew that this report would provoke a more detailed investigation. An ABCC team would come to visit the family and would request information to complete a "long form" questionnaire.[27] Reporting a stillbirth or neonatal death would also provoke an ABCC request for an autopsy. The midwife, then, in reporting an abnormality, stillbirth, or neonatal death, was adding to the family's stress by attracting the attention of the ABCC. If she delayed the report of a stillbirth or

---

24. Committee on Atomic Casualties, "Minutes," 20 April 1948, NAS.

25. Genetics Follow-Up Program, "Report," 2 September 1948, JVN.

26. Masuo Kodani, "Memorandum for the Record," 22 March 1949, 5/NB, JVN.

27. The ABCC used this long form, or "special birth questionnaire," only for abnormal terminations. The form covered family history of the relevant defect ("Does the father have this defect?" "Does any of the fathers' [sic] relatives have any other defect?"), economic status of the family, and menstrual history of the mother. ABCC investigators later decided that the answers provided to many of these questions were not reliable because the questions were too personal. Much of the information obtained regarding abortions and miscarriages was never used (Neel and Schull 1956a, 9–14). The questionnaire is reproduced in Neel and Schull (1956a, 10–13).

neonatal death by a few days the ABCC would still pay her a (somewhat reduced) fee, but the family would be spared the request for the autopsy.[28] (The
fact that many Japanese mothers did not want their infants to be autopsied led
the ABCC to begin calling the autopsy an "examination.")[29]

Midwives' participation in the ABCC study could therefore be professionally risky. The midwives sometimes told families their cooperation was requested by the "military government." An American ABCC staff member, observing this tactic, concluded that it was "the best way for the midwives to
cooperate with us and keep their good relation with their customers at the same
time." They could "make families consent to the reporting by making them
feel pressure of [the] Military Government. The reporting of malformations,
miscarriages, and stillbirths could be made easy for midwives by this trick."[30]

In addition, ABCC pediatricians conducting home visits sometimes attributed a stillbirth or neonatal death to birth trauma, that is, to a poorly handled
delivery. When the reporting of birth trauma became a problem the American
physicians advised the Japanese pediatricians that in such cases the cause of
death should be "recorded in technical terms only, and these explained verbally
to the midwife only if she asks." The death should be attributed to "malfunction
of the respiratory centres" or atelectasis (inadequate distension of the lungs).
"Stillbirths and immediate neonatal deaths will of course show this condition."
No infant death should be attributed to birth injury, the ABCC directive said,
and in any case the parents should never be informed of such a diagnosis.[31]

More commonly an ABCC physician would contradict the diagnosis of a
midwife. One Hiroshima midwife, for example, complained that the pediatricians should be "more careful with their words and try not to disgrace the
midwives." She had informed a new mother that a cephalhematoma, a blood
tumor, would heal in about six months. But the Japanese doctor representing

28. Some midwives may have intentionally delayed these reports. Certainly the ABCC team
often arrived too late to examine an infant, after the funeral and cremation were over. Occasionally
the team arrived as the casket was being carried out of the home, and in several cases, the too-
compliant family stopped the funeral ceremony and opened the casket for the benefit of the ABCC
pediatrician and his nurse (to their horror). "Memorandum for the Record," 26 February 1949,
5/NB, JVN.

29. This policy is discussed in "Instructions to Genetics Doctors Regarding Reporting of Autopsy Findings," 27 September 1951, 5/VB, JVN: "In all reports to *parents* the word 'autopsy' is
to be avoided and 'examination' used instead." (emphasis in original).

30. Masuo Kodani to Carl Tessmer, 26 August 1948, drawer 19, Genetics—1948 #2, NAS.

31. "Instructions to Genetics Doctors Regarding Reporting of Autopsy Findings," 27 September 1951, 5/VB, JVN.

the ABCC suggested the child would have to be hospitalized and given injections to treat it. The midwife was "completely disgraced" by this incident.[32]

The ABCC clerks also consistently differed from the midwives in their calculations of date of delivery. Because they used a different calculating system, the due dates provided by the ABCC clerk and the midwife for any given pregnancy rarely agreed. This was a source of embarrassment to the midwives, who mentioned it in a meeting with ABCC staff member Maki Hiroshi in 1952. The midwives asked why the ABCC clerk did not simply accept the date calculated by the midwives and reported by the mother. All midwives did not, however, use the same method of calculating the due date. Some followed an older method favored by Yamamoto Setsuko, the midwives association president, and others a method more consistent with that used by the ABCC. Yamamoto, who was present at this meeting, insisted on putting the issue to a vote and demanded that the midwives vote against her method. "In spite of our protests and offers to investigate and convert to any system, Mrs. Yamamoto put the matter to a vote, urging them to vote for us and against her, with a resulting overwhelming majority for our present method. We accepted as gracefully as we could, but guaranteed that in the future our estimated date would be revealed to the mothers with hedging and reservations."[33] Yamamoto thus demonstrated her role as an important ABCC ally who smoothed over disputes, defended ABCC policies in public meetings, and encouraged her midwives to cooperate with the ABCC.[34]

In the fall of 1949 American staff learned that the midwives were dissatisfied with the arrangements made for the cremation of stillborn infants provided to the ABCC for autopsy. The remains were sent to a crematorium that cremated the body at ABCC expense and sent the ashes in an appropriate container to the family. But the crematorium in use, Koseikan, while convenient to the ABCC, was not the crematorium with which the Hiroshima midwives had a contract for the handling of all infant deaths. The midwives wished to extract a promise from the ABCC that all future remains would be sent to the Onaga-maki Shiunkan with which the midwives had had a contract since 1929.[35]

The Americans and the Japanese had rather different priorities in negotiat-

32. "Meeting, Hiroshima Midwives Association," 22 September 1951, 5/BC, JVN.

33. Maki Hiroshi to Duncan McDonald, "Memo: Post-Mortem Examination," 24 October 1952, drawer 19, Genetics—1949–1953 #3a, NAS.

34. Schull describes his own interactions with Yamamoto, and his presence at her death-bed (Schull 1990, 88–89, 159–60).

35. Yamamoto Setsuko to Kitamura Saburo, 27 July 1949, drawer 19, NAS.

ing this dispute. In a meeting at the home of the vice-president of the midwives association, Ozaki Tama, the midwives' leadership explained that the Shiunkan had provided "heroic" assistance to the midwives in the period immediately after the war. Yamamoto said it was "absolutely impossible" for the midwives to "break our contract with the Shiunkan." They characterized the ABCC's actions as "contrary to humanity."[36] The American ABCC staff was slightly baffled. The midwives' comments "seem to imply a pre-existing contract," Tessmer noted. He said it was "of considerable importance in this situation to consider the attitude of the midwives, without necessarily acceding completely to their demands." The Koseikan was closer and more convenient, and the ABCC staff felt that sending the autopsy materials to the Shiunkan would offend the Koseikan. To resolve the problem Tessmer amended ABCC policy and began asking that the family specify a crematorium, thereby relieving the ABCC of responsibility for the decision.[37]

HOME VISITS

When a midwife reported a birth, the ABCC sent a medical team to examine the baby at home. The team consisted of a pediatrician, a nurse, and a driver, usually all Japanese. Occasionally an American pediatrician or even a geneticist would go along as an observer. The ABCC team did not make appointments in advance.[38] The group simply arrived at the home and asked to see the baby. The families were surprised by the ABCC visit if they had not been informed by the clerk at City Hall that an ABCC physician would call.[39] Usually mother and baby were home, however, since Japanese women kept their newborns at home for thirty days after birth.[40]

These home visits were time-consuming and generally perceived by the ABCC staff—both Japanese and American—as unpleasant. For ABCC teams to examine all the newborns in Hiroshima and Nagasaki, jeeps had to be on

36. Yamamoto Setsuko to Kitamura Saburo, 30 August 1949, trans. Saiki Chiyoko, drawer 19, Genetics 1949–1953 #3a, NAS. This letter described a meeting held on 22 August 1949 at the home of Ozaki Tama, vice-president of the association.

37. Carl Tessmer, "Memorandum for the Record," 9 September 1949, drawer 19, NAS.

38. Telephone service was not available to most Japanese homes at this time. Appointments for the nine-month examination were made by bicycle messenger, although no appointments were made for the first at-home visit.

39. Doctors making the home visits mentioned this problem in a meeting on 9 March 1949, "Memorandum for the Record," 22 March 1949, JVN.

40. A basic description of the operation is contained in Neel and Schull (1956a).

the road seven days a week. But ABCC administrators considered the home visits so dreary that they required each pediatrician to be out in the jeeps only half-time, twenty hours per week. Their remaining hours were supposed to be spent in self-education in the ABCC library or in independent research. The ABCC also sponsored lectures on "basic science subjects" two afternoons a week, which the Japanese pediatricians were expected to attend.[41]

One factor that contributed to the unpleasantness of these home visits was the difficulty of locating houses. House numbering in Hiroshima and Nagasaki was erratic, based on the home's year of construction rather than sequential location. Many streets had no names, and even named streets had few street signs. Houses within half a block of a major street on a side street were often given addresses on the major street. Sometimes having an address and a family name was not enough, so physicians asked the midwives to begin recording the occupation of the family in an effort to simplify finding the house. Drivers obtained information from the branch police station, the ration office, or the midwife to find a house. If a police officer volunteered to escort the ABCC team to the home, however, the offer was generally declined, "for fear that if we arrived in the tow of a policeman, it would prove embarrassing to the occupants and entail no end of explanation to their neighbors" (Schull 1990, 63). Each home visit took from forty-five minutes to an hour and a half, including locating the home and examining the newborn. Thus one team could usually make only five or six calls in an eight-hour day.[42]

Once the home was located the pediatrician and nurse faced the task of convincing the new mother to allow them to examine her infant. Some mothers refused. Some did not want the ABCC to examine their infant or were unwilling to have strangers come into their home, perhaps because it was not tidy or not well furnished.[43] According to one American nurse employed by the ABCC, the female nurses accompanied the teams because some mothers were more willing to permit a team that included a female nurse into their home.[44]

41. The Japanese physicians, however, apparently did not spend much time in the ABCC library, where most of the journals were in English. Hardie's 1950 report suggests that the Japanese physicians in fact had only about 25 percent free time to devote to "research" or independent study. George A. Hardie, "Report of a Visit to ABCC in Japan, June–July 1950 [Copy No. 3]," NAS.

42. "Memorandum for the Record," 22 March 1949, JVN. Also, George A. Hardie, "Report of a Visit to ABCC in Japan, June–July 1950 [Copy No. 3]," NAS, 77 ff.

43. Maki Hiroshi to Duncan McDonald, "Memo: Post-Mortem Examination," 24 October 1952, drawer 19, Genetics—1949–1953 #3a, NAS.

44. Louise Cavagnaro, personal communication, November 1991: "The Japanese nurses were extremely valuable in obtaining patient cooperation during the home visits." Cavagnaro directed the nursing staff at the ABCC from 1948 to 1951.

As another incentive for cooperation, the ABCC brought a bar of face soap (usually Ivory) to each new mother. Gentle soaps were in short supply in immediate postwar Japan and skin rashes in infants from harsh soaps were a common problem.[45]

The examination itself could be difficult, particularly in the winter, in poorly heated Japanese homes. The infant could not be fully undressed but had to be examined piecemeal, under thick bundles of clothing. And since most of the infants were normal, the pediatricians were rarely "rewarded" with a significant abnormality (Schull 1990, 67). The home visits were demanding and difficult and produced a great deal of routine information at great effort. Schull considered it a "thankless job" that contributed to the ABCC's difficulties in retaining Japanese physicians.[46]

After January 1950, ABCC physicians had a second chance to examine some of these infants. The ABCC began follow-up examinations of a random sample of infants. Each month, depending on the clinic schedule and personnel availability, between 10 and 30 percent of all infants reaching the age of nine months would be seen. The sample was selected based on the last digit of the ABCC registration number assigned for each pregnancy. In a given month, for example, the clinic might examine all infants (or rather, pregnancies) with registration numbers ending with the numeral 1. Naturally, some of these pregnancies had resulted in stillbirths, and some infants had died in the period between the original home visit and the nine-months examination. In these cases an ABCC staff member visited the home to determine the time and manner of death. In addition, the ABCC called in for the nine-months examinations as many of the babies with reported abnormalities as possible.[47]

Japanese mothers were informed by bicycle messenger that they were to bring their infants to the ABCC facilities on Hijiyama Hill for a thorough examination. An ABCC station wagon (at first) or taxicabs hired by the ABCC picked up the women and their babies for their appointments. This nine-months examination, invariably by an American pediatrician, was a highly valued "well baby" service sometimes requested by mothers who did not fall into the selected group.[48] The doctor queried the mother, through an interpreter,

45. Ray Anderson had proposed a soap and towel package, but apparently the ABCC gave out only soap. "We still don't have my much-wanted soap-towel arrangements for home visits." Ray Anderson to J. V. Neel, 6 August 1948, 6/F, JVN. Also see Schull (1990, 66).

46. Schull quoted in George A. Hardie, "Report of a Visit to ABCC in Japan, June–July 1950 [Copy No. 3]," NAS.

47. The nine-month examinations are described in Neel and Schull (1956a).

48. "One of the mothers of an unregistered child asked particularly that she be seen because she did not want to miss the 9-month examination. (NOTE: A special note will be made of this

about diet, motor development, sleep patterns, and bowel movements. If the mother did not object (many did) a blood sample was taken from both baby and mother to check for anemia and syphilis. The nurse weighed and measured the infant. The pediatrician could assess physical and mental development and reassure the mother that her baby was normal. The ABCC staff, recognizing that this service was popular, often emphasized that the examination was free.

CONCLUSION

The ABCC's handling of disputes with the midwives suggests how American policies could be shaped by Japanese concerns, albeit not always in a predictable way. Conforming to accepted practice at the time, the ABCC paid the midwives for reporting births. It relinquished responsibility for selecting a crematorium when that proved to be a sore point with the midwives. And it stopped providing pregnant women with due dates to avoid any conflict with the midwives' calculations. The ABCC staff was in some ways insensitive to the midwives' perspective, yet it was also flexible enough to adjust policies when they caused problems.

For the subjects—the babies and mothers of Hiroshima, Nagasaki, and Kure—participation in the study had both advantages and disadvantages. The well-baby exam at nine months was popular. ABCC requests for infant autopsies and blood samples were not. Participation may have been affected by the common perception that the ABCC was an agent of the military government. Occasionally a woman did refuse to allow the ABCC team to examine her infant. It is possible that some women who failed to register with the ABCC at City Hall did so intentionally and not merely because (as a small sample later stated) they "forgot" or were "too busy." But participation rates were generally high and, at least in the early years, complaints rare.

The ABCC grew rapidly in 1948 and 1949. The early survey team in the spring of 1947 consisted of Neel and Snell. But by November 1947 the ABCC employed eighty Japanese staff members—drivers, nurses, secretaries, and pediatricians—and Neel was negotiating with the Japanese Finance Ministry for the 4.5 million yen the Japanese government was to contribute for the salaries of these Japanese employees (Occupation policy required Japan to pay the salaries of all Japanese employed by the Americans).[49] By May 1949, the ABCC

---

case, since the unregistered series is not seen at 9 months)." Account of a midwives meeting attached to memo, Maki Hiroshi to Duncan McDonald, "Memo: Post-Mortem Examination," 24 October 1952, drawer 19, Genetics 1949–1953 #3a, NAS.

49. James V. Neel to Richard Brewer, 18 November 1947, 6/J, JVN.

had 50 Allied and 150 Japanese employees; the numbers jumped to 80 Allied and 400 Japanese employees by October 1949; 105 Allied and 600 Japanese employees by February 1950; and 117 Allied and 687 Japanese employees—a total of 804 employees—by May 1950.[50] In one year, the ABCC staff had quadrupled in number. A little more than a year later, in September 1951, the total stood at 1,063 employees, 143 Allied and 920 Japanese personnel.[51] This was the largest number of people ever employed by the ABCC.

In 1951 the ABCC had a motor pool of 110 vehicles, 28 in Nagasaki and 82 in Hiroshima-Kure. Its expensive new laboratory complex, on Hiroshima's Hijiyama Hill, had been completed in December 1950.[52] It had nineteen departments, structured to resemble a university medical school, including pathology, bacteriology, biochemistry, hematology, parasitology, radiology, serology, biometrics, vital statistics, employee health, and public relations.[53]

The ABCC's expenses had expanded proportionately, particularly after construction began on the Hijiyama facilities in 1949. The ABCC's initial funding request, in 1947, was for almost $3 million a year—more than $15 million in 1992 dollars—but the AEC approved budgets of only $1.8 million in 1947 and 1948. This allocation was increased to $3.2 million in 1949 to allow for construction and reduced again to $1.8 million in 1950.[54] That nonetheless represented almost one-quarter of the AEC's overall medical research budget for the year, which slightly exceeded $8 million. It is perhaps not surprising that the AEC Division of Biology and Medicine was interested in the operations of the ABCC to a degree bordering on intrusion.

50. George A. Hardie, "Report of a Visit to ABCC in Japan, June–July 1950 [Copy No. 3]," NAS, 4.

51. Summary of ABCC staff numbers compiled by R. Keith Cannan, undated handwritten notes, R. Keith Cannan file, NAS.

52. Merrill Eisenbud, "Visit to Field Operations of ABCC December 20, 1950 to January 10, 1951," 26 January 1951, drawer 28, Mr. Merril Eisenbud (AEC) Report on ABCC Visit—December 1950—January 1951, NAS, 4. In Nagasaki, the ABCC staff in the summer of 1950 included 137 Japanese and eight Americans. There were two full-time and nine half-time Japanese physicians (seven of the latter did not yet have their Japanese medical license). There were also ten Japanese nurses, thirteen genetics clerks, twenty-eight census clerks, eleven guards, and sixteen drivers for the motor pool. The Americans were two physicians, a business manager, two nurses, a chemist, a secretary, and a medical records supervisor. George A. Hardie, "Report of a Visit to ABCC in Japan, June–July 1950 [Copy No. 3]," NAS, 53 ff.

53. Ernest Goodpasture assessed the effectiveness of this structure in his "Report of a Visit to ABCC in Japan, December 18, 1950 to January 5, 1951," in Goodpasture Report, 3 NAS.

54. ACBM, "Minutes, 18th Meeting," 20 October 1949, RG 326, box 3218, DOE, 38. The AEC biological budget that year totaled $6.1 million. See ACBM, "Minutes," 9 April 1949, RG 326, box 3218, DOE.

The scientific work of the ABCC seemed to be going well. The organization began an informal opthalmology survey in 1949 and had an active study of growth and development in Japanese children exposed to the bomb under way. Twenty to thirty infants were examined each day as part of its large, high-profile genetics study. It had begun studies of pediatric growth and development,[55] radiation cataracts,[56] leukemia,[57] and microcephaly in in utero victims of the bombings.[58] It had arranged for a radiation census to be included with the Japanese national census in 1950, and it had an system of updating these records so that survivors could be tracked if they moved. Significant results were emerging for leukemia, radiation cataracts, and genetic effects—in late 1950 the sex-ratio data indicated a significant effect—and the entire program seemed to be going along so well as to justify continued AEC support for at least another five years.[59]

55. This study was carried out under the direction of William Gruelich, who measured, X-rayed, and weighed children exposed to the atomic bombings to determine if exposure to radiation had affected their rate of growth. William W. Gruelich to John J. Lentz (NRC), "Report on Preliminary Study of Japanese Children at Kure, Hiroshima, Nagasaki and Sasebo," 7 October 1947, 6/B, JVN.

56. The ophthalmology study began in 1949, after the discovery of radiation cataracts in the eyes of a Hiroshima survivor who worked in the cafeteria at ABCC. The story of this worker, Kimura Hatsue, was described in *Life* (1949), "The A-Bomb's Children: Study of Half a Million Japanese Reveals the First Delayed Effects of Atomic Radiation," 59. The scientific report was "Atomic Bomb Cataracts" by D. G. Cogan, S. F. Martin, and S. J. Kimura (1949). See also Cogan et al., "Ophthalmologic Survey of Atomic Bomb Survivors in Japan, 1949" (1950); this was also published as ABCC Technical Report 28–59.

57. Leukemia was a known consequence of occupational exposure to radiation (Henshaw 1944b). The first ABCC publication noting an increase in leukemia in atomic bomb survivors was J. H. Folley, W. Borges, and T. Yamawaki, "Incidence of Leukemia in Survivors of the Atomic Bomb in Hiroshima and Nagasaki, Japan" (1952), which was also published as ABCC Technical Report 30–59.

58. Approximately one thousand persons exposed in utero are included in the ABCC-RERF's ongoing studies of survivors. A 1972 study found that of 487 persons who had been exposed to the bomb's radiation in utero at Hiroshima and Nagasaki, 63 suffered from microcephaly. The condition was most prevalent in those exposed between the third and the seventeenth week of gestation. The results were most striking in Nagasaki, where of nine women heavily exposed during this crucial period of pregnancy, eight bore children with microcephaly (Committee for Compilation 1985, 139–41, esp. 141). The first ABCC report was G. W. Plummer, "Anomalies Occurring in Children Exposed in utero to the Atomic Bomb in Hiroshima" (1952); also published in ABCC Technical Report 25–59. For a first-person account by a mother who bore a microcephalic child, see Nagaoka Chizuno, "A Child Whose Mother Was Pregnant When A-bombed," in Takayama (1973, 143–46).

59. W. C. Davison, "Report on a Visit to the Atomic Bomb Casualty Commission (ABCC) in Japan, 14 March–25 April 1951," in Dr. Davison's Report on Visit to ABCC, NAS.

Instead, in early 1951, the NRC Committee on Atomic Casualties voted to terminate the ABCC, after Shields Warren at the AEC cut the ABCC's budget from the requested $1.8 million to $1 million.[60] The decision to terminate was rescinded a few months later—with no interruption of the work in Japan—but the incident laid bare many of the stresses and strains that had plagued the ABCC since its creation in 1946.

60. The decision to terminate is discussed in the minutes of the Committee on Atomic Casualties, 3 February 1951, NAS. Also see appendix to meeting minutes, Machle to Winternitz, 2 February 1951, NAS.

CHAPTER SIX

Political Survival in Washington

D espite general agreement in Washington that a medical study of the survivors was necessary, many advisers and overseers expressed doubts about the suitability of the Atomic Bomb Casualty Commission to carry out that study or the advisability of either Atomic Energy Commission funding or National Research Council stewardship. As Detlev Bronk observed, "Everyone concerned with the establishment of the ABCC . . . experienced periods of misgiving."[1] In the period from 1947 to 1951, AEC advisers and staff questioned the operations of the ABCC many times. The AEC and the NRC had not seen "eye-to-eye" on the ABCC for much of the organization's existence, and the showdown in 1951 was only the culmination of long-

---

1. Committee on Atomic Casualties, "Minutes," *Bulletin of Atomic Casualties*, 1949, 171 NAS.

standing tensions.[2] From the perspective of many observers and participants, the ABCC was not working: it was not screening enough survivors, not building strong enough bonds with Japanese science, not producing scientific results, or not keeping good people, or it was spending too much money on facilities and equipment.

This chapter explores the debate over the policies and legitimacy of the ABCC from 1949 to 1953, suggesting that the organization survived for a number of reasons unrelated to its potential to produce significant scientific results. One of the most compelling reasons for continuing the study was the perception in Washington that the survivors had to be studied by Americans. Even if the survivors were expected to be a weak data source and even if the program were inadequate, an American study was necessary, from this perspective, to counterbalance the Japanese studies.

THE AEC'S RELATIONSHIP TO THE ABCC

The survival of the ABCC depended almost entirely on the Atomic Energy Commission's continued funding; there was no other likely sponsor for the project. But many AEC staff members and advisers were skeptical about the project. The ABCC was thus subject to a high level of AEC scrutiny, which at times affected employee morale and angered ABCC administrators.

The AEC Advisory Committee on Biology and Medicine (ACBM), was created in the fall of 1947 and met in Washington approximately once a month in its first year and once every two months in subsequent years. It was the group most directly responsible for assessing the AEC's policies regarding the ABCC. The members of this committee, guided by Shields Warren, then-director of the Division of Biology and Medicine, reviewed consultants' reports, recommended funding priorities, and provided the scientific and managerial guidance that shaped AEC directives to the NRC and its Committee on Atomic Casualties. "We have not, as a general rule, regarded it as wise or within the scope of our authority to approve or disapprove specific actions of the Division," acting ACBM chairman Joseph Wearn explained in 1953. Instead the committee sought to help the AEC "formulate a well-balanced and integrated research program" and served as "a nexus between the commission and the medical and biological academic world."[3]

2. James V. Neel to William J. Schull, 23 February 1951, notebook 6, Correspondence I, WJS, AA.

3. Joseph T. Wearn to Lewis Strauss, 20 July 1953, RG 326, Minutes, 38th Meeting of the ACBM, 26–27 June 1953, DOE.

The initial ACBM members were chairman Alan Gregg, director of medical sciences at the Rockefeller Foundation; George Beadle, chair of the division of biology at the California Institute of Technology; Detlev Bronk, professor of biophysics at the University of Pennsylvania; Ernest Goodpasture, pathologist and dean of the school of medicine at Vanderbilt University; A. Baird Hastings, Hamilton Kuhn Professor of physiology at the Harvard Medical School; E. C. Stakman, professor of plant pathology at the University of Minnesota; and Joseph T. Wearn, dean of the school of medicine at Western Reserve University and a physician of internal medicine.[4] Meetings of the advisory committee were generally attended by Warren and occasionally by the AEC chairman (David Lilienthal, and later Lewis Strauss) and other AEC staff members, such as public relations director Edward Trapnell.[5]

The AEC's needs—and the interests of members of the ACBM—had a largely unappreciated influence on American genetics in the postwar period. By 1959, an estimated 15 to 20 percent of the members of the Genetics Society of America were engaged in AEC-supported research or training programs. Theodosius Dobzhansky had a grant to study "the genetic structure of natural populations." L. C. Dunn was studying the "evolutionary forces acting on populations" of the mouse, Bruce Wallace the "genetic structure of populations," and Bentley Glass the effects of radiation in tissue culture. The AEC also funded biological and genetics research by George Beadle, James Crow, Luca Cavalli-Sforza, Milislav Demerec, Earl Green, Alexander Hollaender, E. B. Lewis, Richard Lewontin, Newton Morton, H. J. Muller, J. V. Neel, Tracy Sonneborn, J. Stadler, Curt Stern, A. H. Sturtevant, and many other leading geneticists and biologist in the immediate postwar period.[6]

The Manhattan Engineering District began funding genetics research immediately—MED supported the 1943 University of Rochester project on the genetics of flies (Curt Stern) and mice (Donald Charles)—and the 1946 congressional act transforming MED into the Atomic Energy Commission directed the AEC to develop a research program in biology and medicine. A

4. All titles are drawn from entries in the 1955 *American Men and Women of Science* (American Men and Women of Science 1955).

5. The complete minutes of the meetings of this committee are not all available in DOE archives, because some discussions remain classified. The committee also oversaw the AEC's work on biological warfare—a topic that occasionally comes up in the accessible meeting records—and guided the AEC's massive (and growing) program in biological research through the 1950s. The minutes are archived in the DOE archives in Germantown, Maryland, but they are filled with maddening excisions and sections not transcribed because they were "off the record."

6. For a summary of AEC funding for biological research, see U.S. Atomic Energy Commission, *Genetics Research Program of the Division of Biology and Medicine* (1960), a published report summarizing the program and listing grantees.

Medical Board of Review, created to provide initial guidance to the AEC, rec-ommended in June 1947 that "research and training in all aspects of the appli-cation of atomic energy to medical and biological problems be continued and, where profitable, expanded." The AEC Division of Biology and Medicine was founded shortly after this recommendation. The AEC created significant bio-logical research departments at Oak Ridge (Russell's mouse work, and other work on flies, corn, fungi, bacteriophage, and other organisms) and Brookha-ven (plant genetics and cytology, and the mechanisms of the production of chromosome aberrations).[7] It started with eight contracts for genetics research in 1949. The number increased to fifty-five by 1959; many of these were stud-ies funded for five years or longer. The AEC was also supporting research at the University of Pavia, Italy, the University of Chile, and the Inter-American Institute of Agricultural Sciences in Turrialba, Costa Rica. By 1959, the AEC was spending $3.5 million a year on genetics research.[8]

The largest piece of AEC funding for genetics research went to the Atomic Bomb Casualty Commission. The ABCC, unlike most other projects, was funded not through a typical investigator-initiated grant, but through a contrac-tual arrangement with the NRC. This meant that Shields Warren and other AEC officials took a very direct interest in the details of ABCC operations. The AEC sent several staff members to Japan to assess the program in the early 1950s, including George A. Hardie of the AEC Medical Branch and AEC auditor John Lannan in the summer of 1950; Merril Eisenbud, director of the AEC's Health and Safety Division, in December 1950; AEC Division of Biol-ogy and Medicine deputy director John Bugher in May 1951; and, again, Eise-nbud in November 1954. The recommendations in the reports filed by these AEC employees were specific and detailed, outlining, for example, who should control field operations, where new buildings should be located, what should be done about employee commuting problems, how hiring decisions should be made, which scientific programs should be terminated, and which departments should be abolished.[9] In one case an AEC employee specified that a contract with a particular scientist could not be renewed without explicit AEC ap-proval.[10]

My point is that the AEC Division of Biology and Medicine—essentially

7. Ibid.

8. Ibid.

9. Reports on these visits are in drawer 28, NAS: Mr. Merril Eisenbud (AEC) Report on ABCC Visit—December 1950–January 1951; Dr. George A. Hardie (AEC) Report on ABCC Visit, June–July 1950; and Dr. John C. Bugher (AEC) Report on ABCC Visit, March–May 1951.

10. George A. Hardie, "Report of a Visit to ABCC in Japan, June–July 1950 [Copy No. 3]," drawer 28, NAS.

Shields Warren—took an active interest in the day-to-day administrative and scientific business of the ABCC. In addition, AEC officials sometimes made direct requests of ABCC scientists to publish results immediately or to "refute" reports of radiation effects in the popular press, and in at least one case, an AEC official asked Neel to report directly to the ACBM rather than the NRC Committee on Atomic Casualties.[11] The ABCC was technically under the control of the National Academy of Sciences-National Research Council and its appointed scientific advisory committee, the Committee on Atomic Casualties. But the ABCC was also monitored closely—some felt too closely—by the AEC.

WHY FUND THE ABCC?

Throughout much of the first six years of the ABCC's existence, while hiring was proceeding at a frantic pace, rumors of impending doom circulated regularly among the staff in Japan. This contributed to high turnover and low morale. "I have always been impressed with the frequency with which proposals to discontinue the Nagasaki program keep cropping up in one form or another, and the persistence with which suggestions to close out one or the other section of the program keep coming," Bugher told Grant Taylor in August 1953.[12] Streams of consultants sent by the AEC or the NAS made their way to Hiroshima and Nagasaki—sometimes only a few weeks apart—each generally coming away with a different notion of what the ABCC needed, what it should be, how it should be reformed, and whether it should exist at all.

In 1949, after the ABCC had been operating in Japan for a little more than a year, members of the AEC's advisory committee asked whether "the expense was justified by the results" and whether the AEC should "suspend further major expenditures"—specifically, the construction of facilities in the control city of Kure—until a review could clarify the value of the ABCC. In the summer of 1949, this review was undertaken.

11. Neel strongly protested "Although I am certain that the AEC deserves a full accounting concerning our activities in Japan, I found the request somewhat anomalous. I consider myself consultant in genetics to the Committee on Atomic Casualties, and answerable to them directly rather than to the AEC. As matters now stand the AEC will review the genetics program before I have had the opportunity of going over matters with your committee." James V. Neel to Thomas M. Rivers, 2 November 1950, drawer 19, Genetics 1949–1955, NAS. See also, Neel to Herman Wigodsky at the National Research Council, 28 September 1949, 6/K, JVN, and Neel to Schull, 23 February 1951, notebook 6, Correspondence I, WJS, AA.

12. In drawer 6, Miscellaneous Correspondence outside Japan: 1952–1960, NAS.

John Z. Bowers of the AEC Division of Biology and Medicine went to Japan to assess the program and, upon his return, appeared before the Committee on Atomic Casualties with a wide range of complaints and concerns about the operations in Japan. He said there was a "grave question" as to whether the effort to gather control data was detracting from the effort to collect data about the survivors. He also questioned whether too much emphasis was being placed on "research" projects at the expense of "screening," a distinction he did not precisely define (but which was echoed by later consultants, such as Hardie in 1951).[13] "The question is one of balance in the program, and the ability of the AEC to defend expenditures being made for the ABCC program."[14]

The audience was not sympathetic. As committee members began to express objections to Bowers' comments, Joseph Wearn, a member of the AEC's advisory committee who was attending the meeting, responded by summarizing the objections from the committee's perspective. First, he said, the work had not gone forward as rapidly as it should. The hematologic problem had not been attacked, and the NRC had set its sights too high in terms of personnel requirements. But the "chief concern" of the ACBM, Wearn said, was "the low number of actual screenings, and the emphasis on getting research programs under way"; the ABCC's original plans called for examination of twenty thousand survivors a year, but only one thousand had been examined by the fall of 1949, after eighteen months in the field.[15] This was not a message the NRC committee members and ABCC staff members in attendance wanted to hear. "To state that we ran into a hornet's nest is to put it mildly," Wearn later reported.[16]

Bronk proclaimed that the AEC knew the project was a gamble from the beginning and that "once a decision was made to engage in such work" all

13. George A. Hardie, "Report of a Visit to ABCC in Japan, June–July 1950 [Copy No. 3]," drawer 28, NAS.

14. Bowers was also "concerned as to whether or not the administrative and construction aspects of the work are too far ahead of the scientific portion of the program and considered the possibility that the ABCC may end up with handsome, well-equipped facilities but not enough work to justify their existence." Executive session, Committee on Atomic Casualties, 1949, *Bulletin of Atomic Casualties,* NAS, 168–71.

15. Executive Session, Committee on Atomic Casualties, 1949, *Bulletin of Atomic Casualties,* NAS, 172.

16. The formal minutes of this Committee on Atomic Casualties meeting do not convey this "hornet's nest" response. The minutes are terse and nondescriptive, e.g. "Dr. Rivers stated that the Committee desired specific criticisms." Executive Session, Committee on Atomic Casualties, 17 October 1949, *Bulletin of Atomic Casualties,* NAS, 168–73. Also, ACBM, "Minutes," 20 October 1949, RG 326, box 3218, ACBM, DOE, 24.

involved should be "prepared to continue" regardless of the uncertainties involved. Neel asked the two AEC representatives to indicate which aspects of the ABCC program they considered to be "extraneous research"; Wearn quickly reassured him that "there were no questions concerning the genetics program." M. C. Winternitz of the NRC said the project had "values, other than the discovery of atom bomb effects, in its relationship to world health," to which Bronk responded that problems of world health would not justify AEC funding.

The meeting closed with Wearn, apparently taken aback by the response, absolving his fellow advisers on the ACBM of responsibility for one of the actions that most incensed the group, a recent AEC stop order on construction in the control city of Kure. The committee had only "asked for additional information" about the Kure construction, a request apparently interpreted by Shields Warren as reason to immediately stop work.[17]

When Wearn described this meeting to his fellow AEC advisory committee members a few days later, one attendee wondered if the AEC could get out of the deal. He had "grave apprehensions about carrying on this program to the extent that it is being carried on at the present time." He said he continued to believe that the genetics work was necessary but questioned the value of the other medical studies. Could the ABCC's management structure be changed? Perhaps the Committee on Atomic Casualties and the Advisory Committee on Biology and Medicine should have joint meetings. Perhaps the NRC was the wrong organization to direct the ABCC.[18]

TERMINATION

In 1951, the AEC cut the ABCC budget almost in half, a decision the NRC's advisory committee angrily interpreted as intended to terminate the entire program. Ironically, the AEC decision was partly justified by the recommendations of a consultant sent to Japan by the National Research Council. Both the NRC consultant and the AEC advisers felt that the ABCC was in trouble, but they had different reasons for believing so.

In late 1950, the NRC and the AEC agreed to send representatives to Japan to assess the ABCC and its future. The NRC Committee on Atomic Casualties sent physician Willard Machle. The AEC Advisory Committee on Biology and

17. Executive Session, Committee on Atomic Casualties, 1949, *Bulletin of Atomic Casualties,* NAS, 172–73.

18. ACBM, "Minutes," 25 January 1950, RG 326, box 3218, ACBM DOE.

Medicine sent Ernest Goodpasture, vice-chairman of the committee. And the AEC itself sent Merril Eisenbud, director of the Health and Safety Division. Their trips roughly overlapped, though Eisenbud was in Japan a shorter time than the other two observers. Each filed a separate report. Their conclusions led the AEC to cut the ABCC budget and to recommend termination of the organization by January 1953.

Machle's report, particularly, lent support to the growing sentiment within the AEC that it should eliminate funding for the ABCC entirely in the next one or two years. His conclusions about the program in Japan were shaped by a conviction that American involvement in nearby Korea represented the initial engagement of "World War III." [19] Machle recommended that the NRC review its policy with respect to the ABCC; if the NRC chose to continue the project, he said, it should immediately work to improve the ABCC's relationship with the Japanese government and scientific community, strengthen professional support and guidance of the field operations, recruit more Allied scientists, and improve housing, lab facilities, and independent logistic support. In general, he said, relations with the Japanese universities were "excellent" and relation-ships with patients "fortunate." The policy of carrying out exit interviews was helpful in maintaining cooperation, and turning over diagnostic data to family physicians was also important and helpful.

But he felt that the time had come to rethink the future of the ABCC. "Newer elements such as the national emergency and the beginning of World War III require new definition or affirmation of policy." "Social, scientific and political implications of the enterprise" should, he said, be evaluated, and the NRC should consider the "propriety of maintaining workers and their depen-dents in a foreign theatre in the face of almost certain involvement of Japan in this war." Also to be considered: "The costs of continuing with an adequate effort to include housing, logistics, staffing and the like, all taken in relation to the results to be expected from this type of study as compared to those which may be anticipated from the application of an equivalent effort to other means of study." Specifically, he foresaw a radiation study of American populations after the coming war. "With the likelihood that atomic bombing of the United States may occur in the near future, thought may be given to the abandonment of the present project and the preparation of plans for study of populations elsewhere." [20]

19. Willard Machle to Detlev Bronk, "On Field Operations of the Atomic Bomb Casualty Commission," report dated 3 January 1951, 6/Y, JVN.

20. At the request of the chairman of the National Research Council Machle made a study of the ABCC during a visit from 19 November 1950 to 4 January 1951. Machle was in Hiroshima as a consultant to the NRC when Chinese forces crossed the border into North Korea in late Novem-

Virtually alone among the early analysts of the ABCC, Machle was more enthusiastic about the studies of somatic effects than the studies of genetic effects. He considered the genetics study less important because the findings were expected to be negative, that is, no genetic effects were likely to be demonstrated. But the studies of leukemia should be continued, he said: "In view of the status of the field enterprise in Japan and the basic scientific accomplishments already achieved, it would be the worst form of near-sightedness to give up now." At the same time, he noted, the coming war could make abandonment unavoidable. He therefore detailed how the organization could be dismantled. He estimated that it would cost approximately $500,000, most of that for termination and relocation of American staff. The United States could donate the just-completed facilities on Hijiyama Hill to the Japanese government, he suggested. "Except for automobiles," he noted, "little cash recovery can be anticipated." Machle also laid out several other options. The ABCC could be maintained as it was; it could be realigned with the Japan Science Council rather than the Japanese NIH; or it could be turned over to the Japanese or to the American military.[21] Even designating the project an official mission of the U.S. Department of State could have some advantages, Machle suggested, though there was "a certain lack of logic in disguising what is essentially a research enterprise devoted to purely scientific interests [under] the cover of a diplomatic mission."[22]

Goodpasture similarly outlined options and pointed out problems posed by the outbreak of war. He recommended, however, that the AEC support the ABCC for at least five more years. To "abandon the enterprise at this time when one can begin to foresee tangible results" would be unwise, he said, and curtailing the ABCC to the genetics program alone would be "worse than complete abandonment." He unfortunately suggested that the "most important

---

ber 1950. His report, filed with Detlev Bronk in January 1951, was in part a panicky response to the war. Willard Machle to Detlev Bronk, "On Field Operations of the Atomic Bomb Casualty Commission," report dated 3 January 1951, 6/Y, JVN.

21. Willard Machle to Detlev Bronk, "Report on Field Operations of the Atomic Bomb Casualty Commission," 3 January 1951, 6/Y, JVN. ABCC director Carl Tessmer drew up a "military use plan" in July 1950, which suggested possible use for the ABCC facilities in the coming war, including use as a blood bank or a base for a hospital team, radiologic consultation, or an evacuation team. See "Plan for Use of ABCC Facilities and Personnel in the Event of Military Necessity," drawer 28, Dr. George A. Hardie (AEC) Report on ABCC Visit, June–July 1950, NAS.

22. Regarding establishment as a State Department mission, Machle said that the "possibilities of such an arrangement need to be explored from the point of view of professional standing, freedom from taxation and the like. There are certain advantages to such a position."

result" of the ABCC's existence was its impact on medical education in Japan—a result hardly sufficient to justify continued AEC funding.[23]

For his part, Eisenbud said that the "value of the research becomes either more or less valuable as the probability of full scale war increases." To argue that the results will be of value, he said, "we must be able to demonstrate that the findings of the ABCC will be of major public health significance and are apt to lead to important prophylactic, diagnostic or therapeutic measures by which the American public would benefit in the event of atomic attack upon our communities." He pointed out in his report that the plan to cut the ABCC budget to $1 million—in response to the NRC's budget proposal of $2.5 million—was a de facto order to terminate the ABCC. "The sudden retrenchment in FY1952 will involve so drastic a reorientation of the program that efficient administration of a one million dollar budget will be difficult."[24]

The AEC response to these reports was essentially to order the termination of the ABCC. The ACBM, "after a great deal of deliberation," concluded that the ABCC should "accomplish as much as could be accomplished within the next 18 months," but the program should be "drawn to a conclusion as of 30 June 1952." They "might go along" with supporting the project through January 1953, but in any case everyone should begin to make definite plans for phasing out the ABCC.[25] The committee agreed that Japanese scientists should be encouraged to take up the abandoned work and that another agency, possibly the NIH, should be found to support continued American work on genetics, leukemia, cataracts, and other long-term effects.[26]

Meanwhile Shields Warren informed the NRC that the budget proposed for fiscal year 1952 would be only $1 million, and not the $2.5 or amended $1.8 million requested. Within weeks, the NRC's Committee on Atomic Casualties agreed that the ABCC should "be discontinued because of lack of financial support." The motion was passed on the afternoon of 3 February 1950. Winternitz said that the "importance of ABCC in Japanese-American relationships and its influence on Japanese medicine" were "continually increasing" and "furnish an argument against termination of the program;" perhaps the Department of State or the Department of Defense would take an interest in

23. Ernest W. Goodpasture, "Report of a Visit to ABCC in Japan, December 18, 1950 to January 5, 1951", in Goodpasture Report, NAS.

24. Merril Eisenbud, "Visit to Field Operations of ABCC, December 20, 1950 to January 10, 1951," 26 January 1951, drawer 28, NAS.

25. "Notes on a Meeting regarding Conference between Advisory Committee on Biology and Medicine, AEC, and Representatives of the AEC," 16 January 1951, drawer 27, AEC General Correspondence 1946–1955, NAS.

26. ACBM, "Minutes," 12–13 January 1951, RG 326, box 1217, folder 6, DOE.

funding the ABCC. Stern too favored an appeal to the State Department "before announcement of closing is made in Japan," but Machle said the department, while interested, was not prepared to fund the ABCC. John Lawrence of the Committee on Atomic Casualties suggested that the NRC group should meet with the AEC group, and that greater communication all along might have prevented the present situation.

The discussion of the decision to terminate the ABCC was guided primarily by NRC advisory committee chairman Thomas Rivers's contention that the AEC's proposed budget of $1 million for fiscal year 1952 was "merely a polite way of forcing ABCC to close." "Beyond the decision whether the program is to stay or go," the "lack of funds" mandated termination of the ABCC, Rivers said, though he "regretted the necessity to close it." But clearly many of those in attendance at the meeting where this was discussed, including NRC staff members, ABCC staff such as Taylor and Neel, and members of the advisory committee, were profoundly uncomfortable with this decision.[27] "To allow something like this to fold up just when it is getting well under way and beginning to yield results which I sincerely believe, the world being what it is today, justify the admittedly high cost, is to my way of thinking the sheerest kind of scientific stupidity," Neel told Schull.[28]

A little more than a week later a report appeared in New York and Washington newspapers. Wadsworth Likely, a reporter with the Science Service, broke the story on 12 February, saying that the AEC was "killing the only research program able to tell Americans what the long-range effects of an A-bomb attack would be on physical and mental health."[29] The *Washington Report on the Medical Sciences* blamed Shields Warren personally for the "tragic mistake" of cutting funds to the ABCC, citing the ABCC's "achievements to date in adding to knowledge of leukemia and other malignancies, eye cataracts and radiation's genetic influences."[30]

Suddenly that spring, the "improvement of the military situation" made continuation of the ABCC seem more practical, and all the decisions of January and February 1951 gradually unraveled.[31] Through a series of formal

27. Committee on Atomic Casualties, "Minutes," 3 February 1951, NAS.

28. James V. Neel to William Schull, 23 February 1951, notebook 6, Correspondence I, WJS, AA.

29. A slight exaggeration, since the ABCC was not studying mental health; "Only Study of A-Bomb's Effect on Man Is Halted," *New York World Telegram and Sun,* 12 February 1951. A complete manuscript of Likely's story is in drawer 12, Newspaper Clippings 1950–1957, NAS.

30. Excerpt in drawer 12, Newspaper Clippings 1950–1957, NAS.

31. See Bugher's comments, Committee on Atomic Casualties meeting, 24 September 1951, *Bulletin of Atomic Casualties,* NAS, 266–94, esp. 268.

and informal meetings, the ABCC was reconstructed with even greater AEC input.

CONCLUSION

Throughout the ABCC's existence, its relatively low potential for significant or compelling scientific discovery troubled many observers. Some proposed two intuitively contradictory ideas: first, that the study was scientifically suspect or even worthless, and second, that it might turn out to be extremely useful anyway. Eugene P. Evans and Everett I. Pendergrass, for example, suggested in late 1948 that while the program might at present be "a terrible waste of money, time, and effort," it could provide information of "immense value to coming civilizations." They seemed to think that future generations might discern something of value in the ABCC's mass of data, though they themselves clearly did not.[32]

Following a similar line of argument, Lowell Woodbury, a physician who worked with the ABCC, suggested in December 1955 that the ABCC continue to collect genetic information just in case. "I have an intuitive feeling that we will be making a serious mistake to stop collecting the information on the descendants of atomic bomb survivors," he noted. "Times and attitudes change, and while at the moment this information may not seem worthwhile, at a later date it may be highly worthwhile in connection with some as yet unvisualized study."[33]

In 1951, just as the AEC was canceling funding for the work in Japan, AEC geneticist Max Zelle appealed to Neel to refute a damaging report about to be published by Arnold Grobman. Grobman was a participant in Don Charles's radiation project at the University of Rochester during the war. In a popular book, *Our Atomic Heritage,* he suggested that radiation from industrial exposure and fallout from weapons testing would produce a mutant race of humans in a relatively short period. Extrapolating from the results with mice, Grobman tried to predict the genetic consequences of long-term exposure to low-level radiation for workers at Oak Ridge (Grobman 1951). His conclusions were "such that the Atomic Energy Commission is apparently quite concerned about the repercussions this might have on their labor problems down there."[34]

32. Everett I. Evans and Eugene P. Pendergrass, "Report to Dr. Philip Owen," 30 December 1948, drawer 28, NAS.

33. Ibid. Also, Lowell A. Woodbury to James V. Neel, 5 December 1955, drawer 19, Genetics—1959–1968 #1, NAS.

34. James V. Neel to William Schull, 23 February 1951, notebook 6, Correspondence I, WJS, AA.

Neel was not sympathetic to Grobman's argument—he objected particularly to Grobman's use of mouse data as an exact model for human sensitivity—but neither was he interested in rushing to the defense of the AEC. He had to respond in some way to Zelle's appeal, however, and after "considerable soul-searching," he produced for Zelle a long, formal progress report on the ABCC genetics project. This report the AEC could use as it saw fit in responding to Grobman's report, he noted.[35] He added, however, a short, blunt cover letter to Zelle, suggesting that if the AEC wanted scientific refutation of such claims, it might do well to continue funding the ABCC and its genetics project. The "bogey of the potential genetic effects of the atomic bomb" would "rise to haunt" the AEC for many years to come, Neel predicted, unless the commission "took decisive steps to insure the collection of the necessary data." Those steps were, of course, adequate funding for the project in Japan. "Unless the requisite actions are taken in the very, very near future, what is now an adequately functioning machine is going to fall apart quickly."[36]

In his account of this appeal to Schull, Neel added that the AEC had recently approved a $20,000-a-year grant for the next three years to support Neel's study of the spontaneous mutation rate in man. "It's an interesting sidelight that in the storm which the AEC apparently senses to be brewing following the publication of Grobman's book, they seem to intend to refer to this contract, which actually isn't really negotiated yet, as an evidence of their interest in and concern with the mutation problem."[37] Neel thus recognized that the AEC could use its funding of genetics work—even in the absence of conclusive results—as a form of propaganda, to suggest a commitment to workers at risk of radiation exposure. The political utility of the ABCC—its meaning as a goodwill gesture or a demonstration of the AEC's commitment to worker safety—helped it survive the termination of 1951.

In Japan, too, the ABCC was expected to demonstrate American goodwill, though in very different ways. For many Japanese, the ABCC could have shown its good will by providing medical care to the survivors it studied. The angry public debate over the ABCC's no-treatment policy became a forum in which various parties explored a troubling historical question: Who was to blame for the suffering at Hiroshima and Nagasaki?

35. Ibid.

36. James V. Neel to Max R. Zelle, 20 February 1951, drawer 19, Genetics—1949–1955 #3b, NAS.

37. James V. Neel to William Schull, 23 February 1951, notebook 6, Correspondence I, WJS, AA.

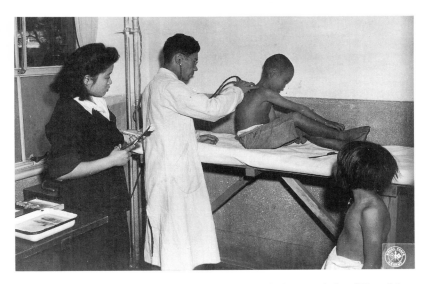

Figure 1. A Japanese physician examining a school child in the control city of Kure, 26 October 1947, at the ABCC clinic maintained in the Japan Mutual Relief Hospital Laboratory. This photo was taken by the U.S. Army Signal Corps. Reproduced with the permission of the National Academy of Sciences.

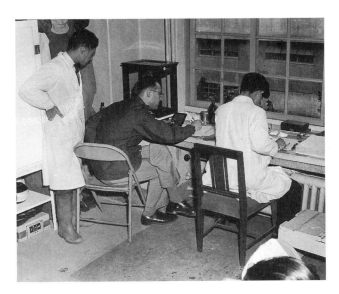

Figure 2. James V. Neel working on the blood studies in the Red Cross Hospital Laboratory of the ABCC in Hiroshima, 26 October 1947. Photograph by U.S. Army Signal Corps. Reproduced with the permission of the National Academy of Sciences.

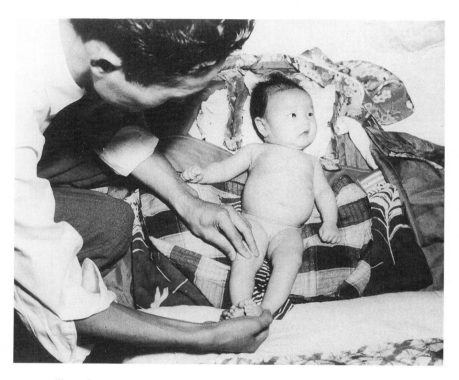

Figure 3. Home visit examination, ABCC Genetics Project, 1948. Reproduced with the permission of the Radiation Effects Research Foundation, Hiroshima.

Figure 4. Louise Cavagnaro, an American nurse and ABCC nursing-staff director, with some of her staff, December 1949. Photo from the George Sakoda Collection. Reproduced with the permission of the Houston Academy of Medicine, Texas Medical Center, Houston.

Figure 5. Technician Velma Yemoto demonstrating pathology laboratory equipment to visiting Japanese physicians, 15 July 1949, at the ABCC headquarters at the Gai-Sen-Kan in Hiroshima. Photo reproduced with the permission of the National Academy of Sciences.

Figure 6. Nurses from hospitals in the city of Hiroshima visited the ABCC's temporary head-quarters at the Gai-Sen-Kan in July 1949. They are shown here in the "animal room." Photo reproduced with the permission of the National Academy of Sciences.

Figure 7. Carl Tessmer, then director of the ABCC, participating in a religious ceremony to purify the ground where new ABCC facilities were to be built. He was placing a green bamboo bough on the altar. Hijiyama, 19 July 1949. Photo reproduced with the permission of the National Academy of Sciences.

Figure 8. Phase 1 construction of the new ABCC facilities on Hiroshima's Hijiyama Hill: view of Unit "E" looking west, showing the completion of galvanized iron roofing, 30 June 1950. The photo was produced by the ABCC photo lab, and is reproduced here with permission from the Archives of the National Academy of Sciences.

Figure 9. Newly completed ABCC headquarters, Hijiyama Hill, Hiroshima, 1951. Photo reproduced here with the permission of the National Academy of Sciences.

Figure 10. Part of the ABCC's fleet of well-marked jeeps, parked in front of its temporary quarters at the Gai-Sen-Kan, Hiroshima, 15 July 1949. The jeeps were used for home visits for the ABCC's genetics program. Photo reproduced with the permission of the National Academy of Sciences.

Figure 11. Jeep #20, Hiroshima, early 1950s. Genetics project home visit teams usually consisted of a Japanese pediatrician, driver, and nurse. Photo reproduced with permission from the Archives of the Houston Academy of Medicine, Texas Medical Center, Houston.

Figure 12. Grant Taylor, who served as director of the ABCC from October 1951 until November 1953, at his desk in 1951. Photo reproduced with the permission of the Radiation Effects Research Foundation, Hiroshima, Japan.

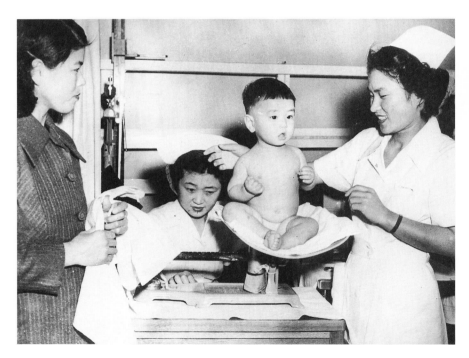

Figure 13. Weighing an infant at the nine-months examination program of the ABCC genetics project. The photo is undated, but nine-months examinations at ABCC facilities were carried out only until 1953. Photo from the William J. Schull Collection, reproduced with permission from the Archives of the Houston Academy of Medicine, Texas Medical Center, Houston.

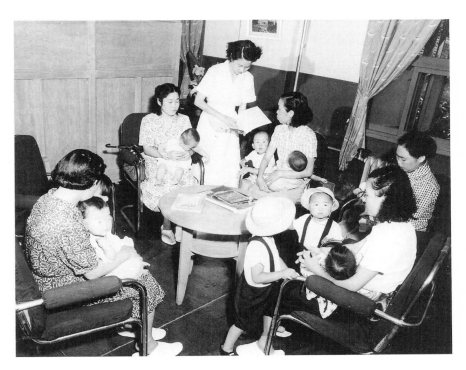

Figure 14. ABCC genetics project waiting room at Nagasaki, early 1950s. Photo reproduced with permission from the Archives of the National Academy of Sciences.

Figure 15. Physician J. Folley, with ABCC nurses at Ujina, Hiroshima, 1950. Photo from the William J. Schull Collection, reproduced with permission from the Archives of the Houston Academy of Medicine, Texas Medical Center, Houston.

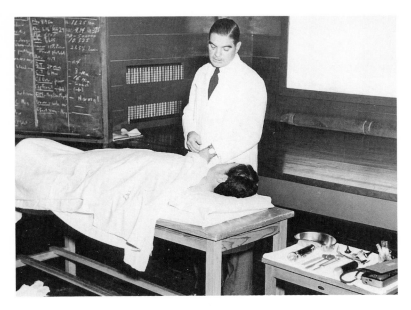

Figure 16. ABCC physician William Moloney with a Japanese patient, at Hiroshima, probably mid-1950s but photo is undated. From the William Moloney Collection, reproduced with the permission of the Houston Academy of Medicine, Texas Medical Center.

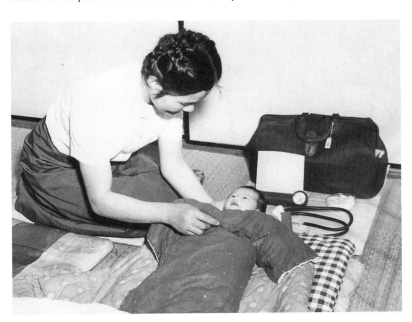

Figure 17. A Japanese nurse on a genetics home visit, 21 March 1950. Photo from the William J. Schull Collection, reproduced with permission from the Archives of the Houston Academy of Medicine, Texas Medical Center, Houston.

Figure 18. Pediatric reception room at the ABCC, Hiroshima, early 1950s. Photo reproduced with the permission of the National Academy of Sciences.

Figure 19. Patient contracting in Hiroshima, January 1952. Reproduced with the permission of the National Academy of Sciences.

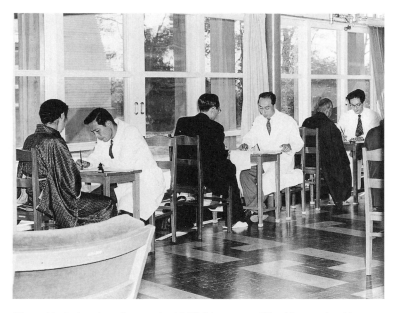

Figure 20. Patient interviews at the ABCC laboratory at Hiroshima, undated but probably mid-1950s. Photo reproduced with the permission of the National Academy of Sciences.

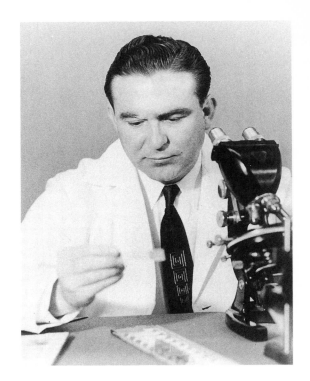

Figure 21. Robert Holmes was director of the ABCC from July 1954 until May 1957. Photo reproduced with the permission of the Radiation Effects Research Foundation, Hiroshima, Japan.

Figure 22. ABCC Assistant Director Hiroshi Maki, probably early 1950s. Photo from the William J. Schull Collection, reproduced with permission from the Archives of the Houston Academy of Medicine, Texas Medical Center, Houston.

Figure 23. George Darling directed the ABCC longer than any other person—from June of 1957 until December 1972. This photo was taken in November of 1967, when he accepted an award from the Japan Medical Association. Photo reproduced with the permission of the National Academy of Sciences.

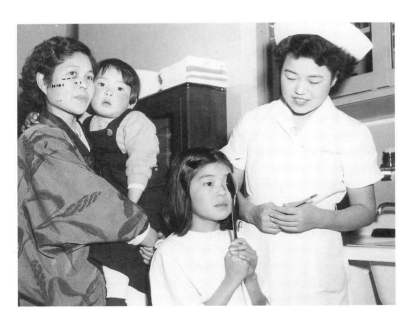

Figure 24. Eye examination in Hiroshima, July 1960. Photo from the William J. Schull Collection, reproduced with the permission of the Houston Academy of Medicine, Texas Medical Center.

Figure 25. James V. Neel, 1965. Photo reproduced with the permission of the American Philosophical Society.

Figure 26. Curt Stern, May 1947, Rochester. Photo courtesy of American Philosophical Society.

CHAPTER SEVEN

The No-Treatment Policy

T he ABCC's medical research required that Japanese survivors come to ABCC laboratories in Hiroshima and Nagasaki for testing and examination on a regular basis. ABCC policy formally precluded providing medical therapy as a compensation for this cooperation. This "no-treatment" policy provoked resentment in the two cities, where the American scientists were commonly perceived as "inhumane" for studying the survivors without taking responsibility for their medical care. Yet in practice the ABCC physicians, both American and Japanese, did provide medical care to both the survivors and the other Japanese residents of the two cities who served as controls. There was a discontinuity between policy and practice.

This chapter explores how those involved with the ABCC—advisers, research subjects, physicians, scientists, and administrators—made sense of the ABCC's no-treatment policy. I suggest that the debate over the treatment policy was shaped by the survivors' unique historical place as the first victims of atomic weaponry. Many Americans interpreted the survivors as war casualties like any others, regardless of their status as the focus of a major American biomedical research project. Providing medical care to these particular casualties, the Americans believed, would amount to publicly "atoning" for the American decision to use atomic weapons. Two questions were thus conflated: First, should the United States atone or apologize for using atomic weapons in Japan? Second, should the United States, through the ABCC, provide medical care to the survivors who were the subject of scientific study? For most ABCC staff and advisers—indeed, for most Americans—the answer to the first question was an unequivocal no.[1] And the answer to the second question followed directly from the answer to the first. The no-treatment policy was grounded fundamentally not in research needs or prevailing medical practice, but in the larger international debate over the legitimacy and morality of the American use of the new weapon in 1945. The United States would not apologize or atone for the use of atomic weapons in Japan, and therefore it would not repair the bodies that had been marked by the bombs' blast and radiation. It would not provide medical treatment to those who had survived the bombings, even though they were the subject of American biomedical research.

My attention to the treatment question reflects the attention it received from ABCC participants at the time. If the no-treatment policy had not been the subject of debate and criticism within the ABCC, it would not now be the subject of historical analysis. Providing medical treatment was not a standard practice in human subjects research in the 1940s. Indeed, it is not a standard practice in contemporary human subjects research.[2] Ethicists continue to de-

---

1. One exception was American anthropologist Earle Reynolds, a pacifist who worked for the ABCC from 1950 to 1953. In 1958 he sailed his yacht through the Pacific proving grounds in the Marshall Islands, in explicit violation of an AEC ruling. Reynolds was arrested and sentenced to two years in jail. He appealed and was never jailed but did become a martyr for the test ban cause. See Reynolds, *The Forbidden Voyage* (1961); Also, Robert Divine, *Blowing on the Wind: The Nuclear Test Ban Debate 1954–1960* (1978, 220–21).

2. Contemporary ethical standards emphasize the importance of taking into account cultural traditions in assessing the propriety of treatment programs. Japan's gift-exchange tradition, for example, would have to be considered in any assessment of the ABCC policy. Robert J. Levine has provided me with this perspective on the problem (personal communication). See also CIOMS, *International Guidelines for Ethical Review of Epidemiological Studies* (1991, 11). "Individuals or communities should not be pressured to participate in a study. However, it can be hard to draw the line between exerting pressure and offering appropriate inducements and creating legitimate motivation. The benefits of a study, such as increased or new knowledge, are proper inducements. However, when communities lack basic health services or money, the prospect of being rewarded by goods, services or cash payments can induce participation. To determine the ethical propriety of such inducements, they must be assessed in the light of traditions of the culture." As Levine has

bate whether medical care should be provided to research subjects, and it is neither appropriate nor possible to state unequivocally that treatment should or should not have been offered under the conditions that existed in Hiroshima and Nagasaki in the early postwar period.[3] This is particularly true in that not all participants in the debate defined "treatment" in the same way. Some limited its meaning to diagnostic services or treatment only of radiation-related illness, others included full medical care for all those participating in the study in any capacity.

The debate, both within and outside of the ABCC, remained heated for at least two decades. Those involved with the ABCC—Japanese subjects, American medical and scientific staff, and administrators and advisers—devoted considerable attention to the question of whether the organization should provide medical care to the survivors. My interest in the treatment question, then, is a consequence of its importance to ABCC participants. Why was medical treatment such a divisive issue?[4] I suggest that the treatment debate was a forum in which various parties explored the proper relationship of the Americans to the Japanese. "Treatment" was a way to talk about the difficulties of international science and about the meaning of the atomic bomb.

---

pointed out to me, given the gift-exchange traditions of Japan, assessing the propriety of material inducements such as free medical care is very complicated. On the problem of providing medical care to "impoverished persons," see Robert J. Levine, *Ethics and Regulation of Clinical Research* (1986, 82–84). See also his discussion of research involving "vulnerable subjects" and how vulnerability can be reduced (1986, 87–93).

3. Assessing the ABCC policy by contemporary ethical standards poses complex problems, beyond the obvious one of imposing contemporary sensibilities on medical researchers in 1948. Ethical review boards in the 1990s sometimes interpret medical treatment of research subjects as "undue inducement" to participate in a study. At the same time review boards have also considered medical care "morally praiseworthy" when investigators are working in communities which lack basic health services (as both Hiroshima and Nagasaki did in the immediate postwar period). There is some evidence that medical care of the survivors and controls by the ABCC would have met a pressing community need. The bomb killed many doctors, nurses, and other health care workers in both cities. In addition, Japanese medical institutions faced serious shortages of supplies and equipment even as late as 1948. In that year Tessmer suggested that the new ABCC building, then in the planning stages, should not be located too close to any Japanese medical center, for "the extreme contrast in adequacy of facilities amounts almost to flaunting such advantages in the face of both Japanese physicians and patients." Carl Tessmer to Everett Evans, 8 December 1948, drawer 28, Drs. Evans and Pendergrass Report on Visit to ABCC—1948, NAS. See also Committee for Compilation (1985, 171). It is perhaps worth noting that there are many variations on this theme: In a recent case, patients with Parkinson's Disease have themselves paid to be included as subjects in a research program using fetal brain cells. See Gina Kolata, "Patients Paying to Be Subjects in Brain Study," *New York Times,* 24 May 1992.

4. Indeed, the debate continued until 1975, when the organization was renamed and reorganized. For a summary of the more extended debate, see Beebe (1979, 204–5).

The ABCC treatment debate was framed by the larger international debate over the legitimacy of the American decision to use atomic weapons in Japan. Some journalists—British and Australian—questioned the morality of American use of the bomb within days of the bombings, and by 1948, when the ABCC began operations in Japan, they had been joined by many others around the world, including influential British physicist P. M. S. Blackett.[5] In his 1948 analysis of the bombings Blackett said that the dropping of the atomic bombs was "not so much the last military act of the second World War, as the first major operation of the cold diplomatic war with Russia now in progress" (1949, 139). Americans had used the bomb, he said, to frighten the Soviets. And he postulated that the discordance between the public explanation ("the advertised objective of saving 'untold numbers' of American lives") and the "realistic objectives" of the bombings, which he said were known or suspected by many, aroused in Americans an "intense inner psychological conflict" (1949, 139).[6]

In the United States, Blackett's hypothesis was immediately "heavily attacked" by both scientists and political leaders.[7] Blackett became (in his own

5. On early coverage of the bombings in the fall of 1945, see Australian journalist Wilfred Burchett's account *Shadows of Hiroshima* (1983). P. M. S. Blackett wrote his study of the atomic bomb as a member of the Advisory Committee on Atomic Energy established by the British government in August 1945. He was trying to develop a "rational basis for policy" regarding "atomic energy" (the term then included atomic weapons) but found that his views differed from those of his colleagues. In the summer of 1947, he began writing an analysis of the "probable influence of weapons of mass destruction on warfare" which he intended to submit to the advisory committee. When the committee was disbanded in the spring of 1948, Blackett expanded his study of atomic weapons and published it in Britain as *Military and Political Consequences of Atomic Energy*. It appeared the following year in the United States under the title *Fear, War and the Bomb: Military and Political Consequences of Atomic Energy* (Blackett 1949).

6. On the extensive debate kicked off by Blackett's interpretation over the next forty years, the two most important historical sources are Gar Alperovitz, *Atomic Diplomacy, Hiroshima and Potsdam: The Use of the Atomic Bomb and the American Confrontation with Soviet Power* (1965), and Martin J. Sherwin, *A World Destroyed: The Atomic Bomb and the Grand Alliance* (1975). Many others tried to dissect the decision to use the bomb, however, including Michael Amrine, *The Great Decision: The Secret History of the Atomic Bomb* (1959); Herbert Feis, *Japan Subdued: The Atomic Bomb and the End of the War in the Pacific* (1961); Len Giovanitti and Fred Freed, *The Decision to Drop the Bomb* (1965); and Walter Smith Schoenberger, *Decision of Destiny* (1970).

7. Paul Boyer discusses the broader post-war debate over the legitimacy of the use of the atomic bomb in *By the Bomb's Early Light: American Thought and Culture at the Dawn of the Atomic Age* (1985, 181–95). On Blackett, in particular, see pages 192–93. Boyer perceptively explores how the bombings at Hiroshima and Nagasaki were interpreted and assessed by editorial writers, political leaders, and the religious press. He notes that the vast majority of Americans expressed approval of the American decision to use atomic weapons at Hiroshima and Nagasaki. At the same time, he notes, some observers as early as 1946 suggested "hidden feelings of guilt"

words) an "atomic heretic."[8] Meanwhile, the decision to use atomic weapons was defended in spectacular terms: "hundreds of thousands of American casualties" prevented (General Leslie R. Groves, who directed the Manhattan Project), avoiding "the ghastly specter of the clash of great land armies" (Secretary of War Henry Stimson) and the "fanaticism" of the Japanese who would "defend their homeland" to the death (physicist Karl Compton).[9] But the violent denunciation of Blackett, and the dramatic language used to justify the bombings as having saved Allied lives, supported Blackett's hypothesis that the bombings were a focus of acute tension in the United States.

President Truman and other American spokesmen blamed the Japanese themselves for the fact that atomic weapons had been used against civilians, arguing that the United States had been forced by "Japan's fanatical leaders" to take drastic action.[10] This construction of the decision attributed moral responsibility for the suffering at Hiroshima to Japanese leaders who refused to surrender. The scenario in which the bomb was dropped to impress the Soviets, in contrast, attributed moral responsibility to American leaders who chose to use an "inhumane" weapon that was not militarily necessary.

My analysis suggests that these stories about the bombings—these conflicting constructions of the historical sequence through which the atomic bombs came to be dropped on Hiroshima and Nagasaki—played an increasingly important role in discouraging a formal treatment policy at the ABCC, even after Japanese demands for treatment began to pose public relations and attrition problems in the mid-1950s. ABCC administrators and scientists cited many impediments to a full-scale ABCC treatment program, including the economic impact on local physicians in Japan, the inability of Americans to qualify for Japanese medical licensure, the idea that treatment was not an appropriate element in a "purely scientific" study, and the expense. Yet the problem of atonement played a role in all these explanatory narratives, since both sides

---

about the bombings that produced in the American people a "vague, ill-defined fear" of future atomic war on American turf (1985, 182).

8. In a 1958 lecture on the BBC Blackett called himself an "atomic heretic" who had, in the intervening ten years, been vindicated. "My military views [regarding the bomb] seem to have become generally accepted. This of course made my crime of 1948 still more grievous, since what can be more tactless than to be right at the wrong time?" (Blackett 1962, 76).

9. All cited in Boyer (1985, 192–93).

10. Truman never expressed any regrets over his decision to use atomic weapons in Japan. In Japanese eyes, he bore "the greatest individual responsibility" for what had happened at Hiroshima and Nagasaki. His perceptions of these events were therefore a source of antagonism in Japan. See Robert Jay Lifton's discussion of Truman in his *Death in Life* (1968, 333–36). See also Bruce Bliven, "The Bomb and the future" *New Republic,* 20 August 1945, 212; and Boyer (1985, 186).

consistently interpreted the survivors not primarily as subjects of medical research, but as political and historical symbols.

Perhaps more than most Americans in the immediate postwar period, those involved with the ABCC were struggling to understand the atomic bomb and its implications. They were engaged in frequent and stressful contact (either direct or indirect) with a group victimized by the bombings at a time when this group was the focus of public sympathy around the world. The survivors were internationally regarded as martyrs of the new atomic age. This status profoundly shaped their interactions with those conducting and managing the research of the ABCC.

GUINEA PIGS

Published criticism of the ABCC was virtually nonexistent during the Occupation, either as a consequence of overt SCAP censorship or because of self-censorship by Japanese journalists unaccustomed to operating as a "free" press.[11] However, when the Allied occupation ended in April 1952, some journalists and authors did begin to question ABCC policies, particularly the no-treatment policy (Imabori 1959, 123–31).[12] Press accounts in Hiroshima began to describe the survivors as "guinea pigs for ABCC," and ABCC critics in Japan and elsewhere seized on the treatment question as evidence of American ill will. "We have truly been faced with many critical issues these past few

11. Braw studied censorship policies of SCAP and her analysis does not entirely back up the suggestion of self-censorship. Her highly informative account of SCAP censorship policies shows how many Japanese authors and journalists were prevented from publishing materials relating to the bombings, since references to the bombings were interpreted in the early years of the Occupation as dangerous to "public tranquility" (Braw 1986). There is some evidence in the ABCC records, however, that local journalists in Hiroshima at least were reluctant to print anything that was not formally approved by American authorities, even in cases in which American authorities told reporters that no approval was necessary. I would argue that self-censorship plays an important role in all journalistic accounts, including those produced under a free press in the United States, and so the situation in Occupied Japan necessarily reflected both SCAP policy and journalistic culture. On reporters' interactions with ABCC staff, see "Memorandum for the Record, Press Conference," 20 December 1948; and "Memorandum for the Record, Interview with Mr. Maekawa, Representative of the Yomiuri Newspaper," 4 March 1949, both in Information Inquiries, Press, Publishers, etc. 1948–1951, NAS.

12. Japanese journalists began to openly condemn the lack of treatment at the ABCC, and one account by a Hiroshima physician went so far as to attribute virtually all anti-American feelings in Hiroshima to the existence of the ABCC and its no-treatment policy (Shigeto Fumio, "The Tragedy of Hiroshima Is Still Living," *Bungei Shunju,* 18 July 1955). For the ABCC's interpretation and response to that article see drawer 19, Genetics—1949–1955 #3b, NAS.

days," ABCC director Grant Taylor told NAS president Detlev Bronk in August 1952. "Labor difficulties, requests for treatment of patients seen by ABCC, attacks from the press and flurries of activities by a group not designated by name. I am sure that what we are experiencing is part of a general wave of anti-Americanism."[13] The same month, a consultant to the NAS, Milton J. Evans, reported that "the pressure is on regarding the need of treatment for A-bomb victims and the Commission [ABCC] is coming in for a share of adverse publicity." Evans said the "actual position" of the ABCC was understood by "fair-minded men in the field of medicine and science in Japan" but the issue was taken up by "those who are unadvised, ill-advised, subversive or anti-American."[14]

Two months later ABCC director Taylor was confronted by a local celebrity, the self-styled "A-Bomb Victim Number One," Kikkawa Kiyoshi. Kikkawa questioned Taylor and Maki Hiroshi, ABCC assistant director, about the treatment policy, saying that he and other survivors were dissatisfied with the lack of medical care at the ABCC. Taylor reportedly responded that survivors were examined and received diagnosis, and since diagnosis is part of treatment, it could be said that the ABCC was providing treatment to the survivors. Kikkawa countered that survivors could not make appointments when they were ill but could only come to the ABCC when asked. Taylor's response was revealing. He told Kikkawa, "I sympathize with you, but you are not the only ones who suffered effects of the war. Therefore there is no cause to render special aid to the citizens of Hiroshima." Taylor thus captured the essential elements of the ABCC explanation in 1952: that diagnosis was a form of treatment and that the survivors could not expect to receive any special consideration, since so many others had suffered in the war.[15] He interpreted the survivors as war

13. Grant Taylor, director of ABCC, to Detlev Bronk at the National Academy of Sciences, 8 August 1952, drawer 3, NAS-ABCC Office, 1950–1953, NAS.

14. Evans included with his report two recent newspaper clippings "which will give you some indication of the direction of the press on the 'treatment issue.'" The articles were from anniversary coverage of the bombing in Hiroshima in the *Asahi Press* and the *Kenzai Shimbum,* both 6 August 1952. See Milton J. Evans to G. D. Meid, 13 August 1952, Atomic Bomb Casualty Commission files, JCB.

15. My account of this meeting between Taylor and Kikkawa is drawn from a later (hostile) published analysis of the ABCC that undoubtedly reflects Kikkawa's perspective. But Taylor's quoted response is remarkably consistent with internal archival materials from meetings in 1955, and reflects, I think, the prevailing ABCC perspective on the treatment question in this period. I am therefore inclined to believe that the published account, whatever its limitations, reflected Taylor's position. Imabori Seiji, "What is A-Bomb Sickness?" official ABCC translation from pages 123–31 of the book *Era of A- and H-Bombs* (Kyoto: Sanichi-Sobo), 21 July 1959, drawer 2, ABCC Press mid-1957–1964, NAS.

victims rather than research subjects and compared them not to other human subjects (such as prisoners or conscientious objectors) but to others who had endured hardships as a consequence of the war.

In 1952 and 1953, public concern about the medical needs of the survivors began to take the form of citizen action groups and medical activism. A group of Hiroshima surgeons, upset at inaction by both the ABCC or the Japanese government, began providing surgery free of charge to survivors in July 1952. (Committee for Compilation 1985, 175–78). The A-Bomb Patient Treatment Councils (A-Bomb Casualty Councils in some translations) formed in early 1953 in Hiroshima and Nagasaki, with the goal of raising funds to provide all needed medical care to survivors, not at the ABCC but at Japanese hospitals.[16] And the A-Bomb Sufferers Association, created by Kikkawa in 1952, tried to arouse public sympathy for the survivors and encourage change in the ABCC policy.[17]

In June 1954 Hiroshima author Kajiyama Toshiyuki published a fictionalized account of life inside the ABCC complex. In *Jikken Toshi* ("Experimental City") the survivors were "guinea pigs," being used by American scientists to help the United States prepare for the next war. The no-treatment policy was presented as evidence of the organization's indifference to the survivors and their needs.[18] The novel's appearance coincided with the public uproar in Japan over the *Lucky Dragon* incident, in which the ABCC and its treatment policy became a special focus of public anger.

The *Fukuryu Maru,* or *Lucky Dragon,* was a Japanese fishing boat dusted with radioactive fallout from the Bravo weapons tests at Bikini atoll in March 1954 (see chapter 8).[19] The twenty-three crew members returned to port suffering from severe radiation sickness. United States authorities were eager to examine these new victims of radiation sickness, and in an unfortunate effort to

16. See report, "Japanese Organizations concerning the Research of Radiation Effects or Treatment of Patients" October 1955, in A-Bomb Survivors Treatment and Welfare 1955–1961, ABCC Collection, NAS.

17. Kikkawa lost control of the organization in 1955 after he was accused of communist ties. See "A-Bomb Sufferers Association Turns Red," translation, August 1955, in A-Bomb Survivors Treatment and Welfare 1955–1961, ABCC Collection, NAS.

18. Toshiyuki Kajiyama, *Jikken Toshi* ("Experimental City"), published in *L'Espoir* (A Japanese magazine with a French name), June 1954; I am indebted to Lifton's analysis of Kajiyama, in *Death in Life* (1968, 413–23).

19. The *Lucky Dragon* incident has been the subject of several detailed accounts, including Ralph Lapp, *The Voyage of the Lucky Dragon* (1958), particularly, in relation to the ABCC, pages 113–15. See also Robert A. Divine, *Blowing on the Wind: The Nuclear Test Ban Debate, 1954–1960* (1978, 4–11), and Richard G. Hewlett and Jack M. Holl, *Atoms for Peace and War, 1953–1961: Eisenhower and the Atomic Energy Commission* (1989, 175–77).

gain access to them, then-ABCC director John S. Morton publicly offered to provide medical treatment for them at the ABCC facilities in Hiroshima. This was "one of the worst mistakes of all" in American responses to the *Lucky Dragon* incident, observed an American anthropologist then living in Japan. "The American government has campaigned for years to convince the Japanese that the ABCC was not set up for treatment and that this was as it should be," Herbert Passin observed in a 1955 analysis of the *Lucky Dragon* incident. "The Japanese have never fully given up their suspicions that they are being used as 'guinea pigs' by the ABCC."[20] The offer to treat the *Lucky Dragon* crew therefore suggested either that the "original contention that the ABCC could not provide treatment was untrue," or that the ABCC "wanted these fresh cases for research purposes."[21] The Japanese scientists overseeing the care of the *Lucky Dragon* victims at a Tokyo hospital declined the ABCC offer (Beatty 1991, 289–91).[22]

Partly as a consequence of Japanese press coverage of the *Lucky Dragon* incident, survivor resentment over the no-treatment policy began to affect rates of cooperation with the ABCC.[23] Patient attrition or family refusal to participate was a "worry for the entire span of the project" but this worry reached "serious proportions" in mid-1954 and 1955. Some ceased to participate because they moved away from the study area, but much of the attrition was a consequence of "reluctance on the part of people to continue to cooperate with the project" because of the "dilemma which people find presented to them when a trip to the ABCC means a loss of a part of a day's wages but no positive contribution to their immediate medical problems."[24] Refusals to participate in

20. Herbert Passin, "Japan and the H-Bomb," *Bulletin of the Atomic Scientists* (October 1955).

21. See Passin (1955); see also Robert Jungk, *Children of the Ashes: The Story of a Rebirth* (1961, 263–74).

22. Beatty notes that a "cold war" broke out between the Japanese and the American medical teams in the course of these discussions (Beatty 1991).

23. Attrition reached 39 percent that month. Thus only 61 percent of those who began the study were still involved. The figure cited in the March 1954 semiannual report of the ABCC, NAS. Many American scientific reports based on the ABCC data include a footnote expressing gratitude to the survivors for their cooperation with the ABCC, and in some cases cooperation was remarkable. Refusal to participate in the genetics program—the effort to detect mutations in the offspring of survivors—was extremely uncommon, "so infrequent that no analysis of the phenomenon has been carried out." But most of the genetics work was carried out during the Occupation (the collection of genetics data ended in January 1954) when relations between ABCC investigators and research subjects were inevitably shaped by the fact of military occupation (Neel and Schull 1956a, 63).

24. Bugher discusses the problem of attrition in "Visit to the Atomic Bomb Casualty Commission Operations, February 1956," GC 1956 200, box 16, file 107, JCB, 11–12.

the ABCC project because of the no-treatment policy continued to be a problem through the 1960s.[25]

The no-treatment policy also fueled the curious idea, popular in Hiroshima, that the two cities had been bombed as part of a biomedical "experiment." Some residents expressed suspicions that the medical studies had been planned in advance and, indeed, that the bombings were undertaken primarily to create a population of heavily irradiated research subjects. (While studies of the bombs' effects on buildings were planned in advance, the biological studies were not.) One journalist asked in 1948 if the story were true that the ABCC planned "to experiment with another atomic bomb, much smaller in scale, of course, to test its effects in Hiroshima again."[26] Such a story drew on an image of scientists as cold-blooded and unfeeling: the ABCC did not treat the survivors, some Japanese critics said, because treatment would interfere with the scientific work, making it impossible for the Americans to document the full course of the diseases caused by radiation exposure.[27] In this climate of suspicion and anger, the no-treatment policy was a serious public relations problem.

It is important to recognize that in practice many ABCC doctors did treat individual patients—sporadically, in "special cases," or when treatment was "experimental." In one celebrated instance in 1953, ABCC physician William Moloney treated a nine-year-old Hiroshima boy, Masaichi Miyamoto, with a promising new drug for leukemia. Aminopterin had successfully produced temporary remissions of leukemia in ten of sixteen children studied at Boston Children's Hospital in 1948. It was one of the first chemotherapeutic agents found to be effective against cancer. Moloney's use of this drug on an ABCC patient received favorable publicity in Japan.[28] In addition, the documentary

25. The Hiroshima Prefectural Ono Home, where many victims exposed in utero with mental retardation and microcephaly were cared for, decided to "refuse the regular examination of the ABCC" because the "ABCC only records symptoms without providing any medical treatment." "A-Bomb Survivors Mental Institute Refuses ABCC Examination," official translation of article in *Mainichi Press,* 24 June 1967, in Japanese Press Clippings 1967, NAS.

26. "Memorandum for the Record," 29 December 1948, drawer 17, Dr. Kobayashi's Press conference, NAS.

27. The idea that the bomb was dropped as part of a biomedical experiment was widely discussed in Hiroshima, even into the 1960s. Lifton explores this belief in his *Death in Life* (1968, 343–54). Jungk in *Children of the Ashes* also examines this question in his assessment of the ABCC (1961, 244).

28. Moloney explained his views about treatment in his diary entry dated 2 September 1953, 101–02. His treatment of Miyamoto with aminopterin was publicized in the *Chugoku Press,* 20 September 1953, MS. Coll. 73, diary of William Moloney, HAM. See also, Farber, Diamond, and Mercer (1948). On the historical importance of aminopterin, see Emil J. Freireich and Noreen A. Lemak, *Milestones in Leukemia Research and Therapy* (1991, 46–47).

record suggests that Japanese physicians employed by the ABCC routinely provided treatment. Young Japanese pediatricians sent out on home visits to inspect newborns for the ABCC genetics study complained in staff meetings that it was difficult to turn down a mother's requests for medical care for her infant. The pediatricians "often encounter cases which could be treated successfully in very simple ways. . . . They feel that it might make it very difficult for them to make a second visit unless they are allowed to do a little more than they can now in their first visit."[29] American staff suspected that the pediatricians did provide such treatment when possible, despite occasional lectures discouraging it.[30]

Some American professional staff members wanted to provide treatment. They valued the clinical experience, and they believed that the promise of treatment would improve participation and therefore data collection. They might also have had professional reasons for favoring a treatment program: American medical boards refused to give physicians credit for the time spent with ABCC, because "the men don't get experience treating patients."[31] By 1955, the ABCC was offering a "diagnostic referral program" in which any Hiroshima resident with a puzzling case could be brought in for referral, regardless of whether he or she was a bomb survivor or a participant in the larger study. One consultant observed that while both American and Japanese physicians had "no significant interest" in the routine examinations of survivors, the physicians were highly interested in these special diagnostic problems. The opportunity to work with such problems, he concluded, was the only thing attracting young Japanese physicians to work for the ABCC. "It is quite clear that it is this phase of the program which aroused the major professional interest."[32]

29. Masuo Kodani discussed the awkward situation for Japanese pediatricians in his memo, Masuo Kodani to Carl Tessmer, 26 August 1948, drawer 19, Genetics—1948 #2, NAS. Also see "Memorandum for the record," 22 March 1949, in Conference with Doctors of the Medical Follow-Up Team of the Genetics Program in Hiroshima, JVN.

30. James V. Neel, interview by author, 20 November 1988, Ann Arbor, Michigan.

31. The single exception was the American Board of Pediatrics, which agreed to give credit "for practice only" in 1950. Grant Taylor was making the appeals and said "one of the main objections" was the lack of treatment. See George A Hardie, "Report of a Visit to ABCC in Japan, June-July 1950 [Copy No. 3]," NAS, 24.

32. Such patients often had "interesting medical problems" and the sole purpose of the program was to "maintain a medical atmosphere" and provide "teaching material" for the young Japanese physicians. Charles H. Burnett to R. Keith Cannan, "Survey of ABCC from October 20th through November 9th, 1955," in Burnett Report, NAS, 2. Since many of these serious diagnostic problems involved patients who were nearing death, and since the ABCC asked that the bodies of such patients be donated for autopsy, "A-bomb victim #1" Kikkawa interpreted the program as a "corpse production factory" where only hopeless cases were examined (Jungk 1961, 274).

Some ABCC administrators, too, favored a treatment policy, most decisively Robert Holmes, who served as director from July 1954 to 1957. Holmes, who personally placed great emphasis on the ABCC's public image, made no ethical arguments in favor of a treatment program. Rather he considered a treatment policy relevant to scientific needs and public relations, particularly in Hiroshima.[33] Not all ABCC staff members were sympathetic to this view. Physician Jack Lewis protested Holmes's plans, and suggested that treatment would be "a bottomless pit financially to my government and a chaotic influence on the economy of the Japanese community."[34]

As a practical matter, no clear line could be drawn between "treatment" and the diagnostic work carried out at the facilities in Hiroshima and Nagasaki. Diagnosis is a necessary part of treatment, and by detecting the presence of a specific problem and informing the patient's physician, the ABCC staff did provide a valuable form of medical care.

In effect, then, the no-treatment policy was more important as policy than as practice. The ABCC maintained the formal policy, but the staff in Hiroshima interpreted it loosely, provided occasional chemotherapy, and overlooked the actions of individual physicians who chose to ignore the restrictions. ABCC administrators even encouraged publicity about special cases of treatment, and spokesmen in both Washington and Hiroshima often promoted the idea that diagnosis was a form of treatment and that therefore the ABCC did provide "treatment." The ABCC did not provide treatment in the sense that its harshest critics would have preferred, that is, drug therapy or surgery for all participants in the study as needed. But the medical staff did treat many individual patients on an ad hoc basis. Since the official no-treatment policy created problems for the ABCC, and since it was not strictly followed in practice, what purposes did it serve?

33. Holmes's views are outlined in "Semi-Annual Report, ABCC, 1 July–3 December 1954," NAS, 9. See also, Jack J. Lewis to Grant Taylor, 2 August 1954, GT. Holmes expanded on his treatment proposal in a letter to R. Keith Cannan at the National Research Council, 10 February 1955, JVN. He recommended that "a hospital of 100 beds be established in Hiroshima and one of 50 beds in Nagasaki, in order to attract and concentrate in one place all the heavily exposed survivors and controls for the purpose of clinical examination, treatment when necessary, whether surgical or medical, and autopsy when death intervenes."

34. Jack J. Lewis to Grant Taylor, 2 August 1954, GT.

EXPLAINING THE POLICY

The question of whether the ABCC should provide treatment to the survivors was explored within the first few months of the organization's existence. An early adviser, geneticist Donald Charles, suggested in June 1947 that treatment would be a suitable compensation to those who participated as research subjects.[35] And James Neel, in a letter to an NAS colleague in January 1948, casually assumed that medical treatment would be a part of the program and suggested that this treatment could be carried out by Japanese employees of the ABCC, who would be licensed to practice medicine in Japan.[36] By the following summer, however, when the treatment question was raised by a Hiroshima physician in a community meeting called by the ABCC, a no-treatment policy was apparently already in effect.

At that meeting, new ABCC director Carl Tessmer explained the policy in ways that were basically consistent with the explanations that appeared off and on over the next decade. The "question of medical treatment for the survivors of the bombs was again brought up" by a member of the audience who "somehow created the impression that the Japanese doctors felt that this would be a more worthwhile contribution to the survivors of Hiroshima." Tessmer explained that the "primary purpose of the ABCC was medical investigation" and that this task was "so large that our efforts and resources would have to be kept in this one direction. As much as aid in other ways appeared desirable, our duties would have to be confined to such medical investigative work as we were discussing with them today. An additional point of not interfering with their practice or relations with their patients was made."[37] Tessmer said that free treatment of the survivors by the ABCC would threaten the livelihood of Japanese physicians in Hiroshima and Nagasaki.[38]

Throughout the 1950s, ABCC officials continued to suggest that the ABCC policy promoted the economic security of local physicians, and Japa-

---

35. At an important advisory meeting for geneticists in June 1947, Charles suggested that some incentive for long-term Japanese cooperation was necessary and thought the incentive might be medical treatment of the participants. "Summary of Proceedings, Conference on Genetics," 24 June 1947, 5/BB, JVN.

36. James V. Neel to Philip Owen, 8 January 1948, 6/J, JVN.

37. See Carl Tessmer, "ABCC Memorandum for the Record," 1 July 1948, drawer 19, Genetics—1948 #2, NAS. See also Japanese Advisory Council for the ABCC, "Minutes," 14 November 1956, box 9, ABCC 1956, file 106, JCB.

38. This was the "chaotic influence on the economy of the Japanese community" to which physician Lewis referred in his letter in 1954. Jack J. Lewis to Grant Taylor, 2 August 1954, GT.

nese physicians, who were presumably being protected, continued to urge the ABCC to offer treatment, citing the pressing needs of the survivors and the importance of treatment to the public acceptance of the ABCC and its program.[39] Fumio Shigeto, a Hiroshima survivor, physician, and director of the Red Cross Hospital, said in 1955 that the competition argument "may be based too much upon a capitalistic formula."[40] In November 1955, the president of the Hiroshima Prefectural Medical Association—who can at least be considered a relevant representative of the Hiroshima medical community—also asked the ABCC to offer treatment. Physician Masaoka Akira told ABCC officials that "the best way to interest patients selected to take part in the ABCC program and improve the sentiment of citizens is ABCC's offering treatment to patients in some form."[41] And Kunio Kawaishi, another Hiroshima physician, wondered in 1956 whether the ABCC concern about local physicians was "sincerely due to the appreciation of professional ethics on the part of the ABCC or whether it is merely an excuse."[42] My point is that regardless of the sincerity of the American claim, it was not an argument that some of those being protected found convincing.

Another early explanation for the no-treatment policy was that treatment would have violated Occupation policy. The Public Health and Welfare Section of the Supreme Commander of the Allied Powers, under the leadership of Colonel Crawford Sams, did specifically prohibit treatment of Japanese by American physicians. The ABCC was nominally affiliated with the Medical Services Division of the Public Health and Welfare Section (General Headquarters

39. At least one Hiroshima physician urged the ABCC to offer treatment as early as 1947. Takehiko Kikuchi was a member of the medical survey team working in Hiroshima from 1945 to 1947. In this capacity he often treated the survivors, "even for minor headaches and diarrhea." This facilitated patient cooperation, he told ABCC officials. Japanese Advisory Council, "Minutes," 14 November 1956, box 9, ABCC 1956, file 106, JCB. One prominent Japanese scientist complained later to an international congress on radiology that the "American scientists working in the atom-scarred city [Hiroshima] have refused since the war to cooperate with Japanese doctors in the treatment of nuclear explosion victims." Reported in "Americans Refuse to Help: Japanese," *National Guardian* (1 September 1962). ABCC staff physician Howard Hamilton, in a memo to George Darling, expressed his distress over this report. See Howard B. Hamilton to George B. Darling, Dr. Maki, and Dr. Nakaidzumi, 3 October 1962, drawer 2, ABCC Press mid-1957–1964, NAS.

40. Shigeto's article in the August 1955 issue of the magazine *Bungei Shunju* was titled "The Tragedy of Hiroshima Is Still Living"; see official ABCC translation by Joji Kenji, 18 July 1955, drawer 19, Genetics—1949–1955 #3b, NAS.

41. Japanese Advisory Council, "Minutes," 9 November 1955, in Japanese Advisory Council Minutes, NAS, 6.

42. Japanese Advisory Council, "Minutes," 14 November 1956, box 9, ABCC 1956, file 106, JCB.

1948, 1–7).[43] The policy was consistent with a general SCAP attitude that the Japanese should take responsibility for their own needs. Sams told ABCC officials that the organization had "no authority to request examinations, obtain specimens or do operations on Japanese patients."[44] Since the ABCC regularly violated at least two of these prohibitions—by requesting examinations and obtaining specimens—and occasionally provided chemotherapy and other forms of treatment, including emergency hospitalization, the relevance of the SCAP position to the ABCC policy is unclear.[45] It is possible that, when the study began in 1947, SCAP's directive was sufficient to discourage the instigation of a treatment policy. But I have not been able to find any archival records which document how the NAS-NRC or the ABCC staff explained or justified the policy prior to the summer of 1948.

Members of the National Research Council Committee on Atomic Casualties and the Atomic Energy Commission Advisory Committee on Biology and Medicine also sometimes explained the no-treatment policy as a consequence of medical licensing restrictions in Japan. The licensing test, in Japanese, was apparently beyond the capabilities of the ABCC's American staff, none of whom were fluent in Japanese. The majority of physicians who dealt directly with the research subjects were, however, young Japanese physicians, recent graduates of Japan's medical schools, hired by the ABCC in Hiroshima or Nagasaki. In September 1951 the ABCC professional medical staff consisted of seventeen Allied physicians or scientists, four Allied nurses, thirty-nine Japanese physicians, and forty-five Japanese nurses.[46] If licensing alone had been the barrier to a treatment program, the ABCC could have delegated all treatment to the Japanese physicians and nurses, as Neel had suggested in January 1948.[47]

43. Also, re involvement with Japanese NIH, see General Headquarters (1948, 73).

44. Crawford Sams mentioned in "Report, 14–22 June 1947, to the Committee on Atomic Casualties," NAS.

45. The SCAP prohibition of treatment of Japanese by American physicians had been ignored by earlier teams of American physicians working in Japan. Shields Warren, describing his efforts with the first American study team sent in to work with the survivors in 1945 said the Americans were soon "treating [Japanese patients] just like the Japanese were." Shields Warren, oral history interview, by Peter D. Olch 10–11 October 1972, Boston, NLM, 67. Yale physician Averill Liebow reported that his examinations of Japanese patients in 1945 included the provision of multivitamin capsules. "We had been advised by our Japanese colleagues that custom required the physician to give every patient some type of treatment after examining him. Actually, the fee is considered payment for the treatment, not for the diagnosis" (Liebow 1965, 110).

46. Committee on Atomic Casualties, "Minutes," 24 September 1951, Bulletin of Atomic Casualties, NAS, 270.

47. James V. Neel to Philip Owen, 8 January 1948, 6/J, JVN.

One impediment to an ABCC treatment program was undoubtedly the cost; it would have been more expensive to combine research and medical care in a comprehensive study of both survivors and controls. But one 1956 estimate put the cost of a treatment program at approximately $30,000 per year, in a program with a budget that regularly exceeded $1 million. (The figures, adjusted to 1993 dollars, are $163,000 for the treatment program in an annual budget of $5.4 million.)[48] This seemingly low estimate may have been realistic, since the plan called for the use of Japanese medical personnel and hospital beds, and it covered only Hiroshima, where the public relations problem was perceived as more acute.[49] In any case the cost of a treatment program was rarely offered as a public, or even a private, explanation for the policy.

It is possible that ABCC spokesmen were reluctant to raise the question of funding, since the ABCC facilities were, by Hiroshima standards, spectacularly modern and high-tech. The ABCC was commonly perceived as a wealthy organization and its new laboratories in Hijiyama Hill in Hiroshima as a "fishcake palace" with a fleet of huge cars and all the latest scientific technologies.[50] Some impoverished Hiroshima residents were so intimidated by this imposing facility that they borrowed dress clothes for their annual examinations at the ABCC.[51] After a visit to Hiroshima in 1949, Norman Cousins wrote that he

48. These equivalent figures are based on an increase in the Consumer Price Index, calculated by the Bureau of Labor Statistics of the U.S. Department of Labor. The April 1956 CPI was 26.9; the November 1993 CPI was 145.7.

49. The emphasis on public relations in Hiroshima was a reaction to the different fate of the ABCC in the two cities. Public criticism of the ABCC was more intense in Hiroshima, either because the Nagasaki office of ABCC was almost entirely run by Japanese or because the "personality" of the city of Hiroshima has tended to be more militant. The two cities are commonly interpreted as having responded very differently to the bombings, with Nagasaki citizens "accepting their fate" and Hiroshima citizens expressing overt anger. For a popular summary of this view, see "Tale of Two Cities," *Time,* 18 May 1962, 22. ABCC veteran and peace activist Earle Reynolds was angered by this *Time* story, which quoted an unidentified ABCC official as saying of Hiroshima, "This is the only city in the world that advertises its misery." Reynolds's response is summarized in "Time Gives Insulting Account of Hiroshima," *Mainichi Daily News,* 27 May 1962. See also other newspaper clippings relating to this incident in drawer 2, ABCC Press, mid-1957–1964, NAS. On the plan for hospital beds in Hiroshima, see ACBM, "Minutes," 26 November 1956, RG 326, box 3218, DOE, 128–45.

50. The nickname *kamaboko-tei,* or "fishcake palace," derived from the resemblance of the ABCC's rounded buildings to an inexpensive fishcake, *kamaboko.* See my chapter 8, note 2, for more regarding this nickname.

51. "The new clinic facilities of the ABCC were at first considered too elegant by the ordinary Japanese. Many women made trips to beauty salons and donned their best apparel to appear in the clinic. Persons of the poorer classes were known to borrow clothing from their neighbors so they would look presentable" (Matsumoto 1954, 71). Matsumoto also produced a detailed account of

"thought of the millions of dollars being spent by the United States in Hiroshima in the work of the Atomic Bomb Casualty Commission" and wondered why "nothing of those millions goes to treat the victims of the atomic bomb." He described the "strange spectacle of a man suffering from radioactive [sic] sickness getting thousands of dollars' worth of analysis but not one cent of treatment from the Commission" (Cousins 1949, 10, 30). The common belief that the ABCC was lavishly funded may have discouraged ABCC representatives from claiming poverty as an explanation for the no-treatment policy, regardless of the actual relevance of funding to the situation. And if funding were the only or even the most important barrier to a treatment program, why did no one try to raise the money, by appeals either to Congress or to the American public?

I would argue that the most important barrier to a treatment program was not cost, licensure, the risk to Japanese physicians, or the legacy of the SCAP policy. It was rather the American perception of the ABCC as an arena in which the problem of moral responsibility for the suffering caused by the atomic bombs should be addressed. In this construction of the ABCC, treatment of the survivors would constitute public U.S. atonement for the use of atomic weapons.

ATONEMENT

By 1955, the ABCC's no-treatment policy was provoking so many public denunciations of the ABCC—in the Japanese press and in public demonstrations in Hiroshima—that the AEC began to reconsider whether treatment should be offered after all. While technically the AEC was simply the funding agency for the ABCC—program management was delegated to the NAS—in practice the commission's Advisory Committee on Biology and Medicine (ACBM) and its professional staff, under the direction of Shields Warren, significantly influenced ABCC operations. This de facto management of the ABCC made the no-treatment policy an explicit AEC concern.

In the wake of the *Lucky Dragon* incident, Merril Eisenbud, manager of the AEC's New York Operations Office, traveled to Japan in November 1954, partly to assess the treatment question. Eisenbud concluded that "treatment"

---

the ABCC exit interview which explores some of these problems of image and class. See Atomic Bomb Casualty Commission, Department of Contact and Facilitation, *Exit Interviewing of Clinic Patients: A Handbook on Methods and Techniques for Exit Interviews,* published in Hiroshima, Japan, February 1951, drawer 12, Newspaper Clippings and Magazine Articles, 1950–1957, NAS.

meant different things to different observers. Some Japanese wanted the ABCC to meet all the medical needs of survivors, regardless of the relation of those medical needs to radiation exposure. Other critics in the United States, including Norman Cousins (*Saturday Review of Literature* editor and promoter of American plastic surgery for the "Hiroshima Maidens"), novelist Pearl Buck, and former first lady Eleanor Roosevelt, were, he said, "under the impression that a large number of the survivors are suffering from the delayed effects of radiation," when actually "delayed radiation effects are limited to a handful of leukemics for whom very little can be done and to perhaps one hundred cases of cataract for whom very little need be done."[52] While thousands of survivors suffered with keloids—painful, overgrown scar tissue— these were not a specific response to radiation exposure but rather a common consequence of skin injury in some racial groups. Keloids, therefore, did not qualify as radiation effects. Eisenbud acknowledged that "more recently the ABCC staff and some consultants have argued for treatment," but their idea of a treatment program was much more limited, involving only therapies necessary for diagnostic purposes. He agreed that some minimal treatment program might be necessary but brought up the problem of medical licensure, since American physicians could not be expected to pass the Japanese medical exam.[53]

John Bugher, reviewing Eisenbud's report before the ACBM in March 1955, said that "from the beginning the organization there was not permitted to attempt medical therapy of the patients who had been injured by the two bombs." There were several reasons for that initial decision, he said, but "only two that are really significant." First, he said, "to do so would have given confirmation to the anti-American propagandists," who would "insist that such treatment should be an act of atonement for having used the weapons in the first place." A policy of providing treatment would be "completely unacceptable for that reason." Second, "to practice medicine in Japan one has to take a very stiff examination and he has to take it in Japanese." Medical licensing in Japan, then, made treatment by American physicians impractical, he sug-

52. On Cousins, see Rodney Barker, *The Hiroshima Maidens* (1985), and also Sheila K. Johnson's perceptive analysis of the role of gender and race in Cousins's activities in Hiroshima, in her *The Japanese through American Eyes* (1988, 46–50). Eleanor Roosevelt visited the ABCC in June 1953 and engaged in a public debate with ABCC director Grant Taylor, apparently over the ABCC no-treatment policy. See Grant Taylor to Detlev Bronk, 16 July 1953, drawer 3, NAS-ABCC Office 1950–1953, NAS.

53. Merril Eisenbud to John C. Bugher, "Visit to ABCC," 21 December 1954, drawer 28, Mr. Merril Eisenbud (AEC) Report on ABCC Visit November 8–13, 1954, NAS, 4–6.

gested. "So there were two good reasons for our people not trying to engage in clinical practice."[54]

Bugher said the ABCC had not been "refractory" to suggestions that it help the Japanese involved in the study. He noted that the "organization has really contributed quite a lot in the Hiroshima and Nagasaki area in the way of diagnostic services and consultation." But he pointed out the important difference between this limited form of treatment and the kind proposed by ABCC critics: the ABCC had not taken any "direct responsibility" for medical care of the survivors and had no intention of doing so.[55] Bugher noted that Tsuzuki had already received $250,000 in Japanese government funds for the construction of a new hospital in Hiroshima, at which research on the biological effects of radiation on the survivors was expected to be conducted.[56] "If a hospital . . . were to be set up for research in this field, that would give medical care to the patients, if an element of competition were to come up, the ABCC project would almost surely lose out," he warned.[57] He thus acknowledged that the no-treatment policy could affect the collection of scientific data, both by discouraging survivor cooperation and by making the project less attractive to physicians interested in clinical work.

In 1961 two AEC officials, responding to a new book critical of the ABCC, articulated the no-atonement argument in more detail. In *Children of the Ashes,* published in German in 1959 and translated into English in 1961, Robert Jungk said that the ABCC policy was based on American fears of appearing to atone for the use of atomic bombs.[58] The defense of the American decision, by AEC administrative representative George I. Mercer and International Affairs Division chief John T. Napier, basically vindicated Jungk's argument, agreeing that medical treatment of the survivors by the ABCC would constitute U.S. atonement for the use of atomic weapons. In a draft press release, they asserted that to treat the atomic bomb survivors would obligate the United States to provide free medical care to all other persons in Japan who

54. Bugher's report was to the Advisory Committee on Biology and Medicine of the Atomic Energy Commission, at their 11–12 March 1955 meeting. See ACBM, "Minutes," 11–12 March 1955, RG 326, DOE.

55. Ibid.

56. This hospital, the Hiroshima A-Bomb Hospital, was in fact opened in 1956.

57. ACBM, "Minutes," 11–12 March 1955, RG 326, DOE, 107.

58. Jungk recognized that the Japanese government, no less than the American ABCC, had failed to meet the medical needs of the survivors. His sympathies were with the survivors and he questioned both Japanese and American responses to their needs (Jungk 1961, 239–45, 265–74, esp. 271).

were "injured and maimed during World War II." [59] They then turned this argument around. If Japanese survivors accepted medical treatment at United States expense, they said, then all Americans injured in the war with Japan would be entitled to medical care at Japanese expense. Thus American medical treatment of the survivors being studied at Hiroshima and Nagasaki would "impose on" the Japanese government "an obligation with respect to United States personnel injured at Pearl Harbor and elsewhere." [60] The argument was logical if one accepted both the definition of the survivors as victims of atomic weaponry (rather than subjects of biomedical research) and the idea that the ABCC was a proper forum in which to establish responsibility for the bombings. In this construction, whoever provided medical care to the survivors would be accepting moral and historical responsibility for what had happened to them. This explained the American insistence that the Japanese government meet the medical needs of the survivors: providing their care meant admitting that it had caused their pain. [61]

59. They recommended that their statement should appear "on State Department letterhead" since "the charges involve not only the AEC, NAS and the ABCC, but also the Supreme Commander of the Allied Powers." This statement, which was apparently never released to the press, dealt not only with the treatment question, but with Jungk's other criticisms of American scientists: that they had "confiscated" anatomical specimens in the fall of 1945; that SCAP had unfairly consored Japanese scientific publication about bomb effects; and that American scientists had prevented Japanese physicians from examining some of the injured survivors. "Reply to Criticisms of the United States Action in Hiroshima and Nagasaki," 29 August 1961, drawer 4, AEC 1951–1961, NAS. On the problem of specimens shipped back to the United States in the fall of 1945 (and not returned to Japan until 1973) see the records of the Joint Army-Navy Commission, OHA. American censorship of Japanese scientific publication is explored in Monica Braw, *The Atomic Bomb Suppressed* (1986). The idea that Japanese physicians were prevented from examining bomb survivors is suggested in Jungk (1961, 231–33), but unlike the other charges (which at least reflect actual events or policies) is not supported by the published and archival record, since Japanese scientists were involved with both the Joint Commission and the ABCC and since, in addition, several independent Japanese studies of the survivors were under way at the same time.

60. "Reply to Criticisms of the United States Action in Hiroshima and Nagasaki," 29 August 1961, drawer 4, AEC 1951–1961, NAS.

61. The statement went on to say that "of more importance" to the ABCC policy was the "fact that neither the Japanese nor the U.S. wished to have American physicians treating Japanese patients. The tacit understanding of both parties was that the ABCC would not provide clinical services." This was, of course, not true, since many Japanese, both survivors and physicians, did express a desire for medical treatment of the survivors by American physicians. See "Reply to Criticisms of the United States Action in Hiroshima and Nagasaki," 29 August 1961, drawer 4, AEC 1951–1961, NAS.

THE INTERNATIONAL DEBATE

Public criticism of the American decision to drop the bomb appeared as early as 7 August 1945, when the Vatican City newspaper characterized the bomb as a "temptation for posterity" and suggested that the Americans should have followed the example of Leonardo da Vinci, who "destroyed his plans for a submarine because he feared that man would apply it to the ruin of civilization" ("Vatican Deplores Use of Atomic Bomb," *New York Times*, 8 August 1945). The American decision was also questioned in an 8 August 1945 article in the *Times* of London.[62] By 16 August—ten days after the bombing of Hiroshima—British Prime Minister Winston Churchill felt compelled to defend the decision in a speech to the House of Commons: "I am surprised that very worthy people . . . should adopt the position that rather than throw this bomb we should have sacrificed a million American and a quarter of a million British lives in the desperate battles and massacres of an invasion of Japan."[63]

Less than a year after the bombings, in June 1946, Norman Cousins and Thomas K. Finletter stated flatly in the *Saturday Review of Literature* that U.S. government leaders chose to use atomic weapons not to save American lives, the publicly stated reason, but to prevent the Soviets from gaining control of Japan.[64] P. M. S. Blackett published his thesis that the bomb was dropped to impress the Soviets, as the first salvo of the Cold War, in 1948.

John Hersey's *Hiroshima* text, appearing in the *New Yorker* brought individual human victims into the debate, provoking American empathy and guilt.[65] Hersey's account of the courage and suffering of six survivors played a role in transforming the bomb from a necessary (albeit terrifying) weapon to a moral dilemma, "making many readers feel that the creators of the atomic bomb are the world's greatest war criminals."[66] At the same time, Lewis Mumford lashed out at the "genocide" of "total war," saying that the bomb "wrapped up this method of extermination in a neater and possibly cheaper package." A. J. Muste's *Not by Might*, in 1947, labeled Hiroshima a "crime" and a "sin of the most hideous kind." Reinhold Niebuhr said the insistence on "uncondi-

62. *London Times*, 8 August 1945, cited in Blackett (1949, 127–28).

63. Churchill, quoted in a compiled history of Hiroshima and Nagasaki, *The Meaning of Survival: Hiroshima's 36-Year Commitment to Peace* (Chugoku Shimbun 1983, 56).

64. Norman Cousins and Thomas K. Finletter, *Saturday Review of Literature*, 15 June 1946, cited in Blackett (1949, 136–37).

65. Sheila Johnson explores the impact of Hersey's account in her *The Japanese through American Eyes* (1988, 43–46).

66. Film producer Sam Marx to James Franck, T. R. Hogness, and Harold C. Urey, 17 September 1946, cited in Reingold (1985, 239). See also Yavenditti (1974).

tional surrender" that lay behind the dropping of the atomic bomb reflected a "nauseous self-righteousness" (Boyer 1985, 218–19).

I want to suggest that this public debate about the propriety and morality of the American decision to use the bomb was reflected in the debate within the ABCC about whether medical treatment should be provided to the survivors. The Americans emphatically did not want to atone for the use of atomic weapons in Japan. And they believed that the Japanese would interpret medical treatment of the survivors as an official U.S. apology for the bombings.

This concern with avoiding even the appearance of atonement unquestionably surfaced in other ABCC activities. For example, ABCC administrators at first considered employing atomic bomb survivors preferentially but decided that "no great advantage would be gained" and that such a policy "might encourage an air of atonement in the ABCC project" (Matsumoto 1954, 71). And when a number of young Japanese women—the "Hiroshima Maidens"— traveled to the United States for plastic surgery in 1955, ABCC director Grant Taylor expressed concern that the Japanese would interpret such American aid as "atonement for American guilt in dropping the bomb."[67]

The problem of atonement may also have played a role in the failure of any political institution to commit itself to meeting survivor medical needs until 1957. The ABCC refused to provide treatment, and the Japanese government failed to provide treatment or insurance or financial aid for medical care. In April 1952 Japan passed the Law for Relief of the War Wounded and for Survivors of the War Dead. This law included some civilians as "war wounded" (such as mobilized student workers and women volunteers), but it excluded the atomic bomb survivors. Their medical needs were acute, yet they found no institution willing to assist them. Finally, in 1957, Japan passed a law specifically to provide medical care to atomic bomb survivors (Committee for Compilation 1985, 393–409).

Under the A-bomb Victims Medical Care Law survivors qualified for two medical exams per year, one general, one detailed. The national government authorized certain clinics in Hiroshima and Nagasaki to conduct the exams. It paid physicians 287 yen for conducting a general exam and 1,048 yen for a detailed exam.[68] The law had at least one unintended consequence: it inspired the ABCC to take financial responsibility for medical treatment of *both* survivors and controls. ABCC physicians did not necessarily provide the treatment, but the ABCC paid for treatment at other facilities. Technically the ABCC was

67. Grant Taylor to Assistant Secretary of State Walter Robinson, 14 May 1955, Correspondence—ABCC 1955–56, GT.

68. J. C. Clarke, special assistant to the business manager, "Report," 15 February 1965, in A-Bomb Survivors Treatment and Welfare 1962–1966, ABCC collection, NAS.

reimbursed by the Japanese government for examinations of survivors covered under the new law, but the money could be (and later was) used for other purposes. The ABCC was therefore running a treatment program just as AEC officials were proclaiming that such a program was impossible.

In a March 1957 report assessing the impact of the new law on the ABCC, two staff members reported that "informal opinion has been given that ABCC clinical facilities could not qualify as a treatment center since they do not provide all of the services of a general hospital." They added that the ready availability of "another program of examination and treatment may make ABCC's program appear less attractive to the exposed groups and lead to an increase in the already high refusal rate."[69] Then-director George Darling therefore sought and received authorization of the ABCC as an examining center. His motivations were not financial but sociological; he feared that survivors "would have their health status evaluated elsewhere and would therefore be lost to the program." But he and other ABCC administrators were uncomfortable about accepting the Japanese government's reimbursement for the exams. He proposed that the ABCC examine survivors without any reimbursement, but this was not acceptable to local administrators; "The officials of the two cities said they could not accept free examinations." As a compromise, therefore, the ABCC informally agreed to use the income from the fees to help survivors, rather than as part of ABCC operating costs.[70]

The funds were deposited in special bank accounts under the names of Maki Hiroshi, the Japanese associate director of the ABCC in Hiroshima, and Nagai Isamu, his counterpart in Nagasaki. By the end of 1964, the ABCC had received payments in yen equivalent to $40,621 from the Japanese government for medical examinations of survivors, and had spent $27,024 of this on providing for the immediate practical needs of survivors, controls, and on courtesy gifts—consistent with Japanese gift-exchange culture—to various local officials who had been helpful in some way to the ABCC.[71] The ABCC bought pajamas, sheets, blankets, and so on for needy survivors and controls; gifts of

69. The law had been in effect less than a month. "The A-Bomb Sufferers Medical Treatment Law and ABCC's Unified Program," March 1957, Moore and Jablon, 3–5, in A-Bomb Survivors Treatment and Welfare 1955–1961, ABCC Collection, NAS.

70. The reimbursement funds were turned over to the Japanese NIH for administration in a separate fund intended solely to help survivors in need. J. C. Clarke, "Report," 15 February 1965, 2, in A-Bomb Survivors Treatment and Welfare 1962–1966, ABCC collection, NAS.

71. The entire process was explained in detail in a letter by NAS business manager G. D. Meid to Samuel L. Hack, assistant manager for research operations of the New York Operations Office of the Atomic Energy Commission. See draft of letter, Meid to Hack, 22 April 1965, and final form of letter (with very minor changes), 1 September 1965, in A-Bomb Survivors Treatment and Welfare 1962–1966, ABCC collection NAS.

milk for anemic patients or those with young children; a day's wages for study participants who complained about losing income in order to visit the ABCC; and medical care or even medical insurance premiums. In one case the ABCC provided 2,425 yen for surgery and 5,000 yen for health insurance for a man who was unemployed and in desperate need of health care.[72] The "Survivors' Fund" bought medical care for needy survivors, donations to the A-Bomb Casualty Councils, in-house purchases that helped the survivors (a television set for the ABCC patient ward), and honoraria for various groups helpful to the ABCC.[73] But the informal system collapsed when an effort was made to "bring the ABSMTL income under orthodox accounting requirements for contract funds." The Atomic Energy Commission, specifically, felt that the fees should be counted as operating revenue, thereby relieving some of the financial pressure on the AEC.[74]

For my purposes, the existence of this "Survivors' Fund," through which the ABCC provided survivors with financial support when they needed medical care, suggests how important a *public* no-treatment policy was to the ABCC. ABCC spokesmen in Washington continued to defend the no-treatment policy even as the organization was using reimbursement funds to purchase medical care for survivors in need.

CONCLUSION

The ABCC research differed from most human subjects research in the immediate postwar period in its international dimensions and its focus on the passive

72. All these examples are drawn from a short sample of "ABSMTL Disbursements" for April, May, June, and July 1964, compiled and translated for inclusion in Meid's letter to the AEC official who was questioning the arrangement; G. D. Meid to Samuel L. Hack, 22 April 1965, 1 September 1965, in A-Bomb Survivors Treatment and Welfare 1962–1966, ABCC collection, NAS. Most of the records of the allocation of the ABSMTL fees were in Japanese, and Meid had only a small sample translated. The case summaries for each expenditure explained why particular gifts (of clothing or bedding, for example) were needed. The list also included payment of overtime to workers at the crematorium; gifts of soap to city workers who had played some role in facilitating the ABCC's work; small cash gifts to teachers at a school where contacting work had been carried out; and assorted bottles of juice to an outgoing director of a crematorium. Such small gifts were a standard part of doing business in Japan, but "in the United States the need to 'buy' the cooperation [of local authorities] by gifts might well be considered unethical. In the mores of Japan, however, this custom is not only acceptable but an obligatory courtesy of well-intentioned, honest businessmen. To do otherwise would probably be construed as ill-will." Ibid., 10.

73. J. C. Clarke, "Report," 15 February 1965, in A-Bomb Survivors Treatment and Welfare 1962–1966, ABCC collection, NAS.

74. Ibid.

observation of a population with unique characteristics.[75] Most human subjects research involved testing new therapies—drugs—on volunteers. Standards for the handling and treatment of human research subjects were not well defined in the 1950s, as David Rothman's 1991 study suggests. Much of the research promoted by military needs during the 1939–1945 war involved human subjects. The Committee on Medical Research of the newly created federal Office of Scientific Research and Development recommended funding for some six hundred research projects focusing on dysentery, influenza, malaria, treatment of wounds, venereal diseases, and physical hardships. Human subjects were essential to this research, and participants were commonly drawn from state institutions (prisons, orphanages, homes for the mentally retarded, asylums) or from the pool of conscientious objectors.[76]

In the ABCC study the interpretation of the population under study had an important influence on the policy. The issue in Japan was not race or class but political meaning. The people being studied in Hiroshima and Nagasaki were players in a debate about history. Who would take responsibility for the historical event that led to their condition?[77]

The common standards of war, the treatment accorded wounded enemy soldiers or civilians under military occupation, for example, suggest that medical care is not necessarily linked to moral responsibility for an injury: One does not treat enemy soldiers or civilians as atonement for having caused their injuries. Medical treatment is not generally construed by either side as an apology or an admission of guilt. Many German civilians received medical treat-

75. There was at least one other large-scale epidemiological survey under way in which medical treatment was denied to a "special population" of research subjects. However, in this case, the Tuskegee Institute study of syphilis, there was apparently no conflict between policy and practice, and the moral parameters were quite different. See Allan M. Brandt, "Racism and Research: The Case of the Tuskegee Syphilis Study" (1985), 331–43). James H. Jones also explores the importance of medical ideas about race in the Tuskegee study. Jones's analysis focuses on the original decision to withhold treatment, in 1932, from which he suggests that the later decision to withhold penicillin followed logically. See *Bad Blood: The Tuskegee Syphilis Experiment* (1981, 1–15, 113–31).

76. Prison inmates who volunteered for drug trials were praised in newspaper accounts for their sacrifices for their country in time of war. The public accepted the participation of prisoners or asylum inmates in medical research as justified by the common good. After the war, many medical researchers expected this acceptance of human subjects research for the common good to continue unchanged. David Rothman's general thesis is that medical practitioners sought to control the protocols for human subjects research through peer review, but were forced by outside pressures to accept public review of their research practices. Rothman explores attitudes toward human research during the war in his *Strangers at the Bedside: A History of How Law and Ethics Transformed Medical Decision-Making* (1991, 30–50).

77. On the general debate, see Martin Sherwin, *A World Destroyed* (1975).

ment from American military physicians during the Allied occupation of their country. Certainly the situation in Germany was quite different from that in Japan; the country was overrun with refugees, "displaced persons," many with parasites and infectious diseases. These refugees were supposed to be cared for by the United Nations Relief and Rehabilitation Administration (UNRRA), but there were too many for UNRRA to handle. As a consequence, they received medical care from Allied military teams, though much of the treatment was provided in the name of public health.[78] In any case Germans and other refugees receiving such treatment from Allied occupiers were not the focus of medical research by American scientists, so it may be doubly misleading to compare the two situations.

The case of the ABCC no-treatment policy suggests how the practice of biomedical science can be shaped not only by social and political context but also by historical narrative. The policy was part of a historical truth claim based on a certain version of the immediate past. The Americans and the Japanese apparently agreed that to care for the survivors was to take responsibility for their situation. That both sides would construct the treatment problem in this way suggests the unique moral and political location of the atomic bomb and of those who survived it.

While the no-treatment policy was a highly visible subject of popular dissatisfaction with the ABCC, it was not the only point on which Japanese critics questioned ABCC operations. Public relations was a persistent concern, as the ABCC staff and its Washington overseers monitored press coverage, squelched popular rumors, and tried to remove the words "atomic bomb" from the organization's title. The ABCC's public meaning was shaped by its status as a diplomatic outpost in Japan, as a tourist attraction, and as a symbol of Japan's defeat.

---

78. UNRRA had 102 teams in Germany but this was insufficient to the task at hand. Military government medical personnel dusted the refugees with DDT, deloused them, provided first aid for minor injuries, and ran ambulance services. Some Soviet refugees complained in May 1945 that they had been "denied medical care" by the Allied military government (the Supreme Headquarters Allied Expeditionary Force, SHAEF, of the U.S. Army), but the complaint itself suggests that medical care was expected and provided (Ziemke 1975, 285–89; 1984, 57).

CHAPTER EIGHT

The Public Meaning of the ABCC

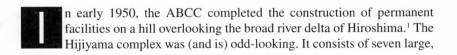n early 1950, the ABCC completed the construction of permanent facilities on a hill overlooking the broad river delta of Hiroshima.[1] The Hijiyama complex was (and is) odd-looking. It consists of seven large,

---

1. The location of the complex on the site of a military cemetery was a negotiated compromise for which Hiroshima's municipal officials were as responsible as ABCC administrators. The city had planned to move the cemetery before the ABCC expressed any interest in the site. Carl Tessmer described the negotiations leading up to the selection of this site in a letter to NRC executive officer Philip Owen, 12 January 1949, 6/H, JVN. Yet the common perception, perpetuated even by some ABCC veterans, was that the cemetery was moved to make way for the ABCC complex (Schull 1990, 29). One Hiroshima physician lamented that the cemetery atop Hijiyama was "dug up and in its place modern smart quonset buildings were erected. Why did they build a research institute on the ground made by digging up the Military Cemetery? Regarding us as guinea pigs, taking advantage of our perseverance, do they even intend to further completely de-

precast ferroconcrete tubes—half cylinders approximately two stories high—within which various offices and laboratories are located. The quonset-style buildings resemble an inexpensive long, rounded fish cake called *kamaboko,* and *kamaboko-tei*—"fishcake palace"—became a derisive term for the ABCC.[2]

The complex also attracted criticism from non-Japanese visitors. In 1951 Merril Eisenbud of the AEC deemed the building's narrow hallways "hazardous," noting that "simultaneous opening of the doors on opposite sides" of a corridor had "already produced one minor injury."[3] John Bugher of the AEC judged the buildings expensive, unattractive, and the product of a deplorable "quonset fixation."[4] Much later, and from a somewhat different perspective, William J. Schull assessed the complex on the hillside as "the most conspicuous symbol in the city of Japan's defeat" and a challenge to the "self-serving rewriting of history" in Japan since 1945 (1990, 29).

The Hijiyama Hill complex was indeed an important symbol in the public debate in Hiroshima over the ABCC.[5] It became both "one of the famous sites

---

prive us of a sense of pride?" Kurokawa Iwao to Robert H. Holmes, 15 July 1955, and Joji Kenji to Robert Holmes, summary translation of "The Tragedy of Hiroshima Is Still Living," 18 July 1955, both in drawer 19, Genetics—1949–1955 #3b, NAS. The "Tragedy of Hiroshima" was an essay by Hiroshima physician Shigeto Fumio which was published in the Japanese *Bungei Shunju* in August 1955.

2. Schull refers to the nickname as *kamabokojo* ("fishcake castle") or *kamaboko-tei* ("fishcake palace") (Schull 1990, 34). Neel (personal communication) says the term was *kamaboko-tei.* Also, "ABCC A Base: Takes Possession of Even the Mind of the Citizens, Survivors Treated as Code Numbers," official translation of article in *Asahi Press,* 5 July 1967, ·in Japanese Press Clippings 1967, NAS. Crawford Sams apparently believed that the permanent facilities were important to public relations, since "as long as the temporary buildings housed the Commission. . . . Japanese physicians didn't feel that the program was there to stay and [the ABCC was] not getting the cooperation from Japanese physicians that was required." ACBM, "Minutes," with comment by John Bowers, 20 October 1949, RG 326, box 3219, ACBM, DOE, 63.

3. Merril Eisenbud, "Visit to Field Operations of the ABCC December 20, 1950 to January 10, 1951," 26 January 1951, drawer 28, Mr. Merril Eisenbud (AEC) Report on ABCC Visit—December 1950–January 1951, Consultant's Visits to ABCC, NAS.

4. Originally the ABCC structures were to be salvage quonsets abandoned in Guam. But after the salvage idea was dropped the "quonset fixation" remained, and special, nonstandard quonsets were built in the United States expressly for the ABCC. Better structures could have been built at lower cost using conventional Japanese materials, according to Bugher. John C. Bugher, "Consultant's Report, Trip to ABCC, March 27–May 7, 1951," drawer 28, NAS.

5. A similar controversy threatened relations in Nagasaki. A January 1950 newspaper story announced that the ABCC was planning to build a "magnificent new laboratory" on the site once occupied by the Tenaka Prison. However, these plans never materialized. The ABCC still wanted to hold onto the option to use the land, possibly for housing, but the governor of the prefecture was eager to have the land released if the ABCC was not going to build the laboratories. George

of Hiroshima included in the course of sightseeing bus routes"[6] and the subject of deep community resentment and anger. In its first two years of existence, it was visited by many foreign journalists, including representatives of the *Berlingski Tidende* (Copenhagen), *National Geographic, Arizona Daily Star, Chicago Tribune, Time* magazine, Sydney (Australia) *Morning Herald, Saturday Evening Post* and London's *Daily Express, Daily Mail,* and *Observer.*[7] In addition, throughout the 1950s, the ABCC complex was visited by American dignitaries, foreign scientists, members of the royal family of Japan—even Marilyn Monroe and her husband, Joe DiMaggio.[8] Japanese visitors included Crown Prince Akihito (twice), Princess Chichibu (twice), actress Yoshiko Yamaguchi (then a member of the Japanese Diet), Princess Takamatsu, and Prince Yoshi. Diplomatic visitors included British ambassador Sir Esler Dening, American ambassador Edwin Reischauer, and former first lady Eleanor Roosevelt.[9] Norman Cousins came by for a meal in the course of his campaign to bring the "Hiroshima Maidens" to the United States for plastic surgery (this annoyed some involved with the ABCC).[10] H. J. Muller came for a scientific tour in 1951.[11] The *General Electric X-Ray News* prepared a story on GE X-ray equipment in use at the ABCC. And *National Geographic* staff members visited.[12] The ABCC was a scientific agency, a minor tourist attraction, a popular stop

---

A. Hardie, "Report of a Visit to ABCC in Japan, June–July 1950 [Copy No. 3]," Consultants' Reports, NAS, 53.

6. "Root of Anti-American Feeling: Dissatisfaction over ABCC," official translation of article in *Asahi Shimbun,* 20 July 1955, drawer 19, Genetics—1949–1955 #3b, NAS.

7. "Memorandum for the Record," 10 July 1949, drawer 12, Information Inquiries, Press, Publishers, etc. 1948–1951, NAS.

8. There is a photo of Monroe and DiMaggio at the ABCC in William Moloney's diary entry for 11 February 1954, MS. Coll. 73, diary of William Moloney, 187, HAM.

9. The Radiation Effects Research Foundation's *A 40th Year Pictorial of Atomic Bomb Casualty Commission and Radiation Effects Research Foundation* (1988) includes photographs of many of these visitors.

10. The AEC's Charles Dunham: "Norman Cousins arrived in town to collect his A-Bomb maidens. . . . [He] seemed to be trying to get as much for nothing as he could. He invited himself and his crowd to dinner the first night he was in town, and proceeded to do it again for breakfast. . . . Dr. Holmes really outdid himself in hospitality for the group, providing free meals and all the facilities of ABCC. All the records on the girls were available, and every one of them, I think, was on the roster of ABCC." See ACBM, "Minutes," 7 May 1955, RG 326, box 3218, DOE, 57.

11. Schull describes Muller's visit (1990, 102–8).

12. David Goodman, editor of *General Electric X-Ray News,* to NRC executive officer Philip Owen, 6 June 1950; also "Memorandum for the Record: Visit of Two Members of the Staff from National Geographic," 7 September 1949, both in Information Inquiries, Press, Publishers, etc. 1948–1951, drawer 12, NAS.

for journalists visiting Hiroshima, and a democratic outpost with some claims to diplomatic attention.

The genetics project was frequently featured in press accounts, with widely varying images of the potential threat posed by the exposure of human populations to radiation. The *Picture Post* ran a story, 6 August 1955, saying that "fifteen thousand [children] are now being born in Hiroshima every year. . . . And all of them are normal." Since children born in Hiroshima and Nagasaki manifested the proportion of abnormal births to be expected in any population, this statement was problematic.[13] At the opposite extreme, in August 1958, the London *New Statesman* ran a graphic description of the bomb's potential genetic effects. Citing figures compiled by the "Japanese Imperial Atom-Research Institute" (according to Neel, no organization by this name existed in Japan), the article said that one in every six children born in Hiroshima was deformed or stillborn (Morris 1958).[14] As the agency in charge of documenting genetic effects, the Atomic Bomb Casualty Commission was the focus of both adulatory and hypercritical reporting, in Japan and around the world.

Public relations was not a minor concern at the ABCC. ABCC staff interacted with the community in unlikely forums—appearing, for example, on an amateur television variety show in Hiroshima to compete with the Hiroshima Medical Society in such skills as juggling and Nō dance.[15] ABCC director Robert Holmes estimated in 1955 that he devoted 90 percent of his time to public relations. He said associate director Maki Hiroshi devoted about 50 percent of his time to the same topic.[16]

13. "The Children are Healthy" 6 August 1955, photocopy, drawer 12, Newspaper Clippings 1950–1957, ABCC, NAS.

14. Neel was responding to an inquiry from CBS News producer George A. Vicas, who wondered if the statistics quoted in the *New Statesman* piece were accurate. "The reason for this letter is to ask for your advice in the matter of the statistics quoted in the article." He wondered if "the Japanese, in their touchiness, might not have lost some scientific objectivity in the release of the figures quoted in the *New Statesman* article." Neel said that he suspected that the report was a "fabrication" and that the Imperial Japanese Atom Research Institute did not exist. "None of a number of very well-informed Japanese and American sources has ever heard of such an institute." See Vicas to Neel, 17 September 1958, and J. V. Neel, undated draft response, drawer 2, ABCC Press mid-1957–1964, ABCC, NAS.

15. And, supposedly, "permitting" the Japanese team to win by purposely flubbing the last juggling trial: "The contest was close, so close that the Hiroshima doctors won by a single point, and wouldn't have won at all if the ABCC's juggler hadn't dropped a ball—at the whispered instigation, some witnesses later suggested, of Dr. Darling, who, it was claimed, thought it would be diplomatic to let the home team triumph." The television program, which occurred in 1959, is described in E. J. Kahn, "Letter from Nagasaki," *New Yorker,* 29 July 1961, 52–54, 54.

16. R. Keith Cannan, "Memorandum of My Visit to ABCC, October 15 to November 10, 1955," drawer 28, Cannan-Hastings Report, NAS.

Press coverage of the ABCC was mixed. Most American accounts, both during and after the Occupation, were enthusiastic. The tone of Japanese reports changed from Occupation-era conciliation and praise to post-Occupation criticism. In the 1950s, ABCC officials blamed the Japanese press for public anxiety about the long-term effects of the bombings, particularly for the widespread concern in Japan about genetic effects. Thus George Darling, who became ABCC director in 1958, charged that "sensational reporting techniques" and "irresponsible treatment" in the Japanese press had increased the anxiety of survivors about their offspring. He said the "very real need of the people to know the truth . . . underlies the attempt at scientific precision and dispassionate objectivity." [17]

But the ABCC itself contributed to public anxiety in different ways. The very qualities ABCC scientists and administrators stressed in their interactions with the Japanese press—the emphases on "rigorous scientific data" and on "reassuring" survivors—may have exacerbated survivors' feelings of victimization. By stressing the "purely scientific" aspects of the organization, scientists at the ABCC thought they were reassuring the populations at Hiroshima and Nagasaki that the results would be trustworthy. But to many Japanese, particularly the residents of Hiroshima, devotion to "pure science" represented an inappropriate, dehumanizing attitude toward the suffering of the survivors. As Lifton has documented, survivors sometimes interpreted "reassurance" about radiation effects as implying that their suffering was not real (1968, 511–16). Japanese journalists and novelists exploited common negative imagery of scientists as cold and unfeeling to depict the organization as inhumane. Rather than insulating the ABCC from political controversy or criticism, then, the "pure science" defense made the organization more vulnerable to Japanese attack.

THE ABCC IN PRESS ACCOUNTS

In the 1950s, coverage of the ABCC and the genetics project in the U.S. mass media was overwhelmingly positive. Reports packaging and promoting the organization often received prior editorial approval from the Atomic Energy Commission. In 1955 the NBC "March of Medicine" television show hailed the "exciting prospects" of atomic medicine at the ABCC, and *U.S. News and World Report* produced a series of articles on the "healthy, happy" babies of

---

17. In a reprint of Darling's 6 August 1958 letter to the *Japan Times*. Darling, a Yale professor of human ecology, was then the new director of the ABCC. A copy of the pamphlet is in box 32, Atomic Bomb Casualty Commission correspondence, Thomas Francis papers, BHL.

Hiroshima and Nagasaki. Associated Press and the *New York Times* wrote stories focusing on the genetics studies as the most important negative finding of the ABCC ("Hiroshima After-effects Mild," Associated Press, 29 March 1955; Robert Trumbull, "Atom Survivors Usually Normal," *New York Times,* 31 May 1955).

Much of this American reporting was so reassuring as to be inaccurate. *U.S. News and World Report,* for example, headlined its exuberant interview with ABCC director Robert Holmes "Thousands of Babies, No A-Bomb Effects" (Report on Hiroshima 1955). The story included such photo captions as "No upsurge of cancer cases" (though the ABCC had already documented high leukemia rates) and "Their children are normal. Radiation burns have healed. There's no radiation blindness" (though radiation cataracts had been identified and some patients lost their vision). Holmes managed to aggravate both H. J. Muller and the NRC with his breezy dismissals of concerns about genetic effects.[18] By December 1955 NRC executive officer Philip Owen recommended "on principle" that "Holmes be instructed that he desist from any further comments on the genetics of the bomb in any way, shape or manner from now on."[19]

But Holmes's comments, while misleading, conformed to the biases of the American press. Reporters often downplayed long-term radiation effects as enthusiastically as AEC spokesmen, even raising irrelevant issues about which to reassure readers. When Robert P. Martin, a regional editor who had witnessed bomb tests at Bikini, interviewed Holmes, he asked if radiation had affected the personalities of the children of survivors. There was no reason to expect personality deviations in the offspring of survivors. In any case, by 1955, the oldest child represented in the ABCC genetics data would have been seven years old, but probably would not have been seen by the ABCC for about six years. Furthermore, none of the questions asked by the ABCC dealt with personality development. The ABCC had no reason to believe that radiation exposure in a parent would affect the personality of the offspring, was not

18. See Beatty (1991, 304–5) on Holmes and the NRC response. Holmes was apparently mystified by the complaints about his comments, since his wording was almost identical to that in a Neel and Schull manuscript describing the genetics results.

19. Owen shared his recommendation by telephone with Neel, and Neel in turn shared with Owen a memo that H. J. Muller was circulating to various geneticists. Muller's memo said that the "irresponsible and misleading statement attributed to Dr. Holmes" made him suspect that the NAS should "restrict the making of such statements by persons speaking officially for the ABCC." Such "seemingly official" statements did a great deal of damage. See Philip S. Owen, "Memorandum for the Record," 15 December 1955, "Telephone Call from Dr. Neel re ABCC." Also, Philip S. Owen to Keith Cannan, 15 December 1955, and Owen to Neel, 15 December 1955, all in drawer 19, Genetics 1949–1955 ABCC, NAS.

studying genetic effects on personality, and therefore had no pertinent information. Holmes nonetheless answered that the ABCC had not noticed any personality defects in the children of exposed parents, and Martin took this as good news.[20]

Few American journalists interviewed Japanese survivors or depicted the ABCC from the perspective of those subjected to study. An unlikely exception was Bill Hosokawa, an American nisei who wrote for the *Denver Post*. Hosokawa was notorious among Japanese-Americans for his endorsement of the wartime concentration camps in the American west. Imprisoned himself in Wyoming, Hosokawa edited the camp newspaper and from this platform denounced camp "troublemakers" and urged his fellow prisoners to cooperate with the War Relocation Authority (Drinnon 1987, 286–87). Yet when Hosokawa came to Hiroshima after the war his account focused on the perceptions and worries of a single Hiroshima family, a husband and wife (both exposed) and their two sons, one of whom had been exposed in utero.[21] Though Hosokawa concluded by noting these survivors' pride at making "a sizable contribution to the world's scientific knowledge," he at least recognized the complexity of the relationship between the survivors and the American researchers (Hosokawa 1950).[22]

Many other American journalists simply interviewed the American scientists and portrayed the ABCC's relationship to the Japanese community as one of warm cooperation. This relationship was in fact characterized by considerable ambivalence. While survivors had some things to gain by cooperating with the ABCC, some expressed mixed feelings about that cooperation. The sense that the survivors, already affected by the American bombs, were being victimized a second time by the American scientists was frequently conveyed in the popular press in Japan and elsewhere. Whether a large proportion of the survivors themselves felt victimized by the medical studies is unclear. In many cases, those charging the Americans with using the Japanese as "guinea pigs" were not themselves survivors. Lifton points out that novelists and journalists took up the cause of the survivors in ways that did not necessarily reflect survivors' interests (Lifton 1968, 413–32).

20. Robert P. Martin, "Report from Hiroshima: Latest about After-Effects of A-Bomb," *U.S. News and World Report,* 13 May 1955, photocopy in Newspaper Clippings 1950–1957, ABCC, NAS.

21. It is unclear whether Hosokawa submitted his manuscript to the AEC prior to publication. Report on Hosokawa's visit, in Public Relations Department, "Monthly Activities Report, October 1950," drawer 12, Information Inquiries, Press, Publishers, etc. 1948–1951, ABCC, NAS.

22. A copy is filed in Public Relations Department, "Monthly Activities Reports," drawer 12, NAS.

Jikken Toshi ("Experimental City"), a fictional account of life inside the Hijiyama Hill complex by Toshiyuki Kajiyama (who was not an atomic bomb survivor), portrayed the ABCC's scientists as callous, cold-blooded, racist, and fanatical. Kajiyama's characters included a toady nisei and a corrupted (Americanized) Japanese woman, both employed by the ABCC (Kajiyama 1954).[23] Hiroyuki Agawa's *Ma No Isan* ("Devil's Heritage"), a 1957 novel also examined relations between the Japanese and the Americans at the ABCC. Agawa's more restrained account depicted the American organization as both terrible and useful; his characters both praised and questioned the ABCC. Thus a Japanese mother in the ABCC waiting room responded to a visitor who asked about the American examination of her baby: "Well . . . how shall I put it? . . . It's wonderful to have the children receive such careful examinations but . . . " (Agawa 1957, 15). And the de facto segregation of some ABCC facilities was briefly mentioned when a visitor was shown around: " 'Was that the dining room we saw when we went down the corridor?' 'Yes.' He wondered a moment why they had not gone there. 'Do the American and Japanese doctors eat together in the dining room?' Miss Kimura was silent a moment and then answered, 'Separately' " (1957, 27). Another character said, "I work here at the ABCC every day, but what bothers me most is that the Americans don't seem to show any visible sign of regret that the thing was dropped on Japanese civilians who did not have the power to fight." (1957, 213). This literary portrayal of the ABCC—as both problematic and helpful—reflected coverage of the organization in the Japanese press. But compared to many press accounts in this period, Agawa's prose was mild and his criticisms balanced.

If the selection of press translations preserved in ABCC archival records can be taken as representative, Japanese press coverage during the Occupation tended to be positive. Japanese newspaper accounts sometimes mentioned such popular concerns as the lack of treatment, but reporters were likely to dismiss these concerns, noting, for example, that the ABCC did not provide treatment because it was only "conducting research." And when ABCC associate director Maki Hiroshi refused to comment on the possible genetic effects of the bomb, one reporter interpreted this refusal as "not due to any political or military secrecy but [the result of] a purely scholarly attitude."[24]

During the Occupation, the "American" character of the ABCC was a source of wonder and praise rather than, as it became later, resentment. A 1950

23. For my discussion of Kajiyama's *Experimental City,* I am indebted to Lifton (1968, 413–23, esp. 415, 419).

24. "A Visit to the ABCC," *Yukan Asahi Press,* 3 March 1950. The translation of the article by Peggy Green is filed with Public Relations Department, "Monthly Activities Report, March 1950," drawer 12, Information Inquiries, Press, Publishers, etc. 1948–1951, ABCC, NAS.

report in the *Yukan Asahi Press,* for example, marveled over the fancy American equipment and the high status of nisei doctors at the ABCC.[25] Accounts of press conferences with Japanese journalists suggest their initial interest in cooperating with the ABCC during the Occupation. At one press conference, several Japanese journalists asked exactly "what could be printed" of what they had been told. They were under the impression that the ABCC was carrying out secret research and "that they might be questioned if they used their own discretion in reporting the information." Lieutenant Colonel Carl Tessmer, then ABCC director, assured them that this was not the case, but added that if they were concerned about accuracy the ABCC would be happy to look over articles in advance.[26]

A zealous willingness to accommodate American censorship was also evident. The Japanese press was accustomed to censorship—a "free press" was a novelty in Japan—and Japanese journalism continues, even in the 1990s, to emphasize consensus and downplay conflict. From the Meiji restoration in 1868 on, government codes controlled Japanese journalists. Police were vested with wide-ranging powers and could interpret censorship laws idiosyncratically. Violators were frequently imprisoned. While specific press codes changed many times between 1900 and 1941, all included censorship. After December 1941, nothing at all could be published in Japan without government permission.[27]

The censorship policy imposed by SCAP perpetuated and reinforced self-censorship among post-surrender Japanese journalists. In one case, a representative of *Yomiuri Shimbum,* one of the major Japanese national newspapers, sought desperately to get prior permission from the ABCC to produce a report on atomic energy. Tessmer assured him that "such permission was not the con-

25. "The stereo x-ray apparatus is rare in Japan. . . . The microtome slices study tissue to a thinness of 4 microns. If all the new and interesting equipment is to be mentioned there will be no end to it. Each patient changes into a freshly laundered gown and is escorted by a Japanese nurse through the various rooms where necessary data is taken." Also: "It is interesting to note that within this organization we can see a picture of the advancement in the scientific fields made by the Niseis which is specially noteworthy in America after the war." Two of the Japanese Americans, the article said, "destiny returned to their parents' native town, Hiroshima." But "neither of these two men made comments on the disaster of Hiroshima." From "A Visit to the ABCC," *Yukan Asahi Press,* 3 March 1950, translation filed with Public Relations Department, "Monthly Activities Report, March 1950," drawer 12, Information Inquiries, Press, Publishers, etc. 1948–1951 ABCC, NAS.

26. Carl Tessmer, "Memorandum for the Record," 20 December 1948, drawer 12, NAS.

27. On Japanese censorship policies, see Lawrence W. Beer, *Freedom of Expression in Japan: A Study in Comparative Law, Politics, and Society* (1984). See also Braw, *The Atomic Bomb Suppressed* (1986, 23, 61–88).

cern or function of the ABCC," but the reporter was apparently "not convinced" and continued to request a written statement of approval.[28] At a press conference in December 1948, several reporters told Tessmer that because of censorship they were reluctant to print anything at all about the ABCC.[29] This reluctance may have been as problematic as the later flood of criticism. Rumors about the ABCC flourished in a climate in which little information was available to the public.[30]

Though formal censorship of the Japanese press ended in 1949, Japanese reporting on the ABCC did not change significantly until after the Occupation ended in 1952. The ABCC was among the victims of the anti-Americanism that appeared in Japan in the summer of 1952.[31] By the mid 1950s, reporters for the nationwide newspapers, *Mainichi Shimbum* and *Asahi Shimbum,* and the local newspapers, such as the *Chugoku Press* in Hiroshima, began to play up reports by Japanese scientists that seemed to contradict the results of the ABCC, particularly reports of genetic effects. Japanese journalists openly criticized the lack of treatment at the ABCC, and one account by a Hiroshima physician went so far as to attribute virtually all anti-American feeling in Hiroshima to the existence of the ABCC.[32] By the late 1950s, the privileges enjoyed by the "foreigners" at the ABCC began to be depicted as unfair, and even standard laboratory procedures—the thin slicing of tissue specimens for study, for example—were labeled "inhumane."[33]

The ABCC staff responded to this reportage in several ways. The staff publicly discounted some Japanese scientific accounts of the biological effects of radiation but simply ignored others. The response apparently depended on how much publicity the Japanese scientist had attracted. When a Nagasaki University physician reported that 20 percent of all survivors' children were

28. Carl Tessmer, "Memorandum for the Record," 4 March 1949, drawer 12, NAS.

29. Tessmer, "Memorandum for the Record—Press Conference," 20 December 1948, drawer 12, NAS.

30. One journalist, for example, asked in 1948 if the story was true that the ABCC planned "to experiment with another atomic bomb, much smaller in scale, of course, to test its effects in Hiroshima again." "Memorandum for the Record," 29 December 1948, drawer 17, Dr. Kobayashi's Press Conference, NAS.

31. Kawai Kazuo, *Japan's American Interlude* (1960).

32. Shigeto Fumio, "The Tragedy of Hiroshima Is Still Living," *Bungei Shunju,* August 1955. For the ABCC's interpretation and response to that article see drawer 19, Genetics—1949–1955 #3b, NAS.

33. "Foreigners under Special Status without Registration," *Chugoku Press,* 10 March 1959, and "The Actual State of ABCC," *Chugoku Press,* 7 August 1959, both in drawer 12, NAS.

abnormal, the story was widely quoted, even in the United States.[34] The physician, Okamoto Naomsa, claimed his work was done in collaboration with the Nagasaki ABCC, which particularly incensed then-director Robert Holmes. The ABCC registered a legitimate protest with the dean of the university, Kitamura Seiichi, asking that Okamoto "re-examine the integrity of his statement and so-called scientific findings."[35] But the ABCC staff was more likely to ignore reports of genetic effects unveiled at Japanese scientific meetings if they did not attract publicity.

In at least one case, the ABCC director suggested that a Japanese report of genetic effects might be correct. In early September 1949, two Japanese newspapers in Hiroshima carried a report that a taller Japanese race would result from the genetic effects of radiation on the survivors, an effect some might consider positive. ABCC director Carl Tessmer issued a "special statement" allowing that such a report was plausible: "It is well known that such changes as those mentioned in the newspaper article are a possibility in any organism exposed to radiation, [and] such conjectures may be a valuable part of a scientific approach." Deformity was also a possible effect, of course, though the ABCC staff did not, to my knowledge, ever describe conjectures about monstrous babies as a valuable part of a scientific approach.[36]

Responding to attacks on the ABCC itself was somewhat more complicated, particularly in Hiroshima. Journalists and authors have often portrayed Hiroshima and Nagasaki as responding differently to the experiences of the atomic bombings. Hiroshima's monuments to the bombing emphasize Japanese victimization more explicitly than the modest monument at the Nagasaki hypocenter. To many Americans Nagasaki appeared to have handled the experience more appropriately. An 18 May 1962 *Time* article, for example, depicted Hiroshima as "grimly obsessed by the long-ago mushroom cloud," while Nagasaki "lives resolutely in the present."[37]

The ABCC's experiences bore out the stereotypes. Hiroshima always

---

34. "A-Parents Have Deformed Babies," *Nippon Times,* 5 June 1956, and "Genetic Effects of Radiation?" *Chugoku Press,* 4 June 1956; official translations of both are in drawer 12, NAS.

35. Robert Holmes to Kitamura Seiichi, 19 June 1956, box 9, ABCC April–July 1956, JCB.

36. Public Relations Department, "Monthly Activities Report, August 1949," drawer 12, Information Inquiries, Press, Publishers, etc. 1948–1951, ABCC, NAS.

37. This essay also claimed that the ABCC had not documented any increase in rates of leukemia among bomb survivors, though by 1962 the ABCC had already published data indicating a significantly greater risk of leukemia among heavily exposed survivors (Folley, Borge, and Yamawak: 1952; Moloney and Kastenbaum 1955). The essay managed to annoy residents of both cities. Pacifist Earle Reynolds, an anthropologist and antinuclear activist then teaching at Hiroshima's

posed more serious public relations problems, and in 1949 the ABCC established a public relations department in Hiroshima to help deal with continuing problems. This department distributed the pamphlet "What Is ABCC?" to schools and civic centers, invited student reporters from high school newspapers to tour the ABCC facilities, and organized conferences with principals, school nurses, doctors, and parent-teacher groups. The ABCC later selected fifty leading citizens of Hiroshima, "purely for public relations purposes," to participate in an adult health study. These prominent citizens received regular free medical examinations, under the assumption that this would encourage their support of the ABCC.[38]

As the end of the Occupation loomed, the ABCC also began to focus on improving its image among survivors. This effort included trying to counteract unpublished, sometimes fanciful local rumors about the ABCC by tracking down and "correcting" the individuals who spread them. Numerous rumors were thus squelched in Hiroshima—for example, that semen collected in infertility studies was used for artificial insemination, that genetics doctors gave injections to rid Hiroshima of abnormal babies, and that only impotent males were subjected to testicular biopsy because it would damage the testicles.[39]

The ABCC also established a formal "patient contacting" department to handle "exit interviews." In 1951, Tessmer appointed Scott Matsumoto, an enthusiastic nisei with a B.A. from American University, to direct this department in Hiroshima.[40] Matsumoto was the architect of the ABCC exit interview; his handbook, prepared for ABCC staff members in 1951, suggested the rather complex purposes of these interviews. The interview followed the annual phys-

Women's College (Jogakuin), sent a formal protest to *Time,* as did the mayor of Hiroshima and then-ABCC director George Darling. The Nagasaki chapter of the Mitsubishi Shipbuilding Labor Union protested comments by Nagasaki mayor Tsutomu Tagawa suggesting that Japan would have used atomic weapons if it had developed them. Mary McCracken to George Darling, 25 April 1962, describes Reynolds asking Darling to protest (he did, through a letter published in *Time,* 8 June 1962). Also, *"Time* Gives Insulting Account of Hiroshima," *Mainichi Daily News,* 27 May 1962, and "Repercussion to *Time's* 'Tale of Two cities,'" *Yomiuri Press,* 26 May 1962. All in drawer 2, NAS.

38. Scott Matsumoto, "Patient Rapport in a Foreign Country," a pamphlet produced by the Atomic Bomb Casualty Commission, drawer 12, Newspaper Clippings and Magazine Articles 1950–1957, NAS.

39. For a description of this campaign, see Scott Matsumoto, "Patient Rapport in a Foreign Country," drawer 12, Newspaper Clippings and Magazine Articles 1950–1957, NAS.

40. George A. Hardie, a consultant to the Atomic Energy Commission, described patient contacting as the ABCC's "waste basket," a department that took over "things that other departments don't want to handle." George A. Hardie, "Report of a Visit to ABCC in Japan, June-July 1950 [Copy No. 3]," NAS, 73.

ical, which could take several hours. The physical included rectal examinations, X rays, blood sampling, dilation of the pupils, and in some cases pelvic
examinations and sternal punctures. Regardless of its potential long-term medical benefits for the person examined, this physical was not a pleasant experience. When the patient was finished and dressed, he or she met in a small
private room with an ABCC staff representative, and the interviewer attempted
to defuse any complaints about the ABCC before the subject returned to the
community. "We must be able to gauge what the patient will say to his family,
his neighbors, and his friends about ABCC upon his return home," Matsumoto
said. "His comments in his own community may decide whether others will
cooperate with ABCC or not." [41] The interviewer should "have the patient state
his dissatisfaction within the ABCC," then try to "reverse an unfavorable impression." A patient allowed to voice his objections would be "less likely to
repeat them once he is out of the clinic." [42]

The interview should be "informal," he said, and "any sense of its being
an investigation or cross-examination should be strictly avoided." He also suggested discussing "the weather, the latest baseball score of the Hiroshima
Carps," or other general topics. On sensitive issues, he advised diplomacy. "Do
not say, 'The ABCC does not give treatment,'" Matsumoto said. If the patient
should ask, the interviewer was to explain that the "ABCC renders a thorough
examination to determine the true health condition of the person and if medical
care is indicated, ABCC refers such patients to their family doctor or to the
medical association." This, Matsumoto observed, said the same thing, though
in more palatable language. He also suggested that interviewers tell patients
they were contributing to medical science, helping all of humanity, and in the
process gaining knowledge of their "true medical state." To be a good interviewer required patience and flexibility. The "best contactors" were women
twenty-five to forty years old, he said, though "the better unmarried female
contactors usually left the commission before too long, as the prerequisites for
a good contactor, obviously, were the same as for a good wife." [43]

The ABCC did change several policies in response to patient complaints
raised in these exit interviews. Many patients expressed annoyance at the long

---

41. Scott Matsumoto, *Exit Interviewing of Clinic Patients: A Handbook on Methods and
Techniques for Exit Interviews,* pamphlet issued by the Atomic Bomb Casualty Commis-
sion, Department of Contact and Facilitation (Hiroshima: Atomic Bomb Casualty Commission,
February, 1951), drawer 12, Newspaper Clippings and Magazine Articles, 1950–1957, NAS.
Matsumoto also published a revealing account of his work in the *American Journal of Nursing*
(1954).

42. Ibid., 71.

43. Ibid.

wait for the diagnostic report on their examination, so the ABCC expedited the report schedule. The staff removed tombstones from the "sacred Japanese Army graveyard" around the ABCC after one patient complained about them. The ABCC also started hand-delivering positive findings of syphilis, after patients complained of spouses or other family members opening reports sent through the mail, in an effort "to put patients at ease." [44]

However, the primary purpose of the postexamination interview was to discourage public complaints about the ABCC. I have not been able to find any documentation of its impact on the patient-participants—their impressions of the interview at the end of a long and difficult day were presumably not shared with their questioners—but Matsumoto felt the interviews met the ABCC's needs. In 1954 he published an enthusiastic report on the success of the program and its positive impact on ABCC public relations in Hiroshima (Matsumoto 1954).

A CRISIS IN PUBLIC RELATIONS: THE *LUCKY DRAGON* INCIDENT

On 1 March 1954 twenty-three Japanese fishermen aboard a tuna boat, the *Fukuryu Maru,* or *Lucky Dragon,* encountered radioactive fallout from the Bravo weapons test at Bikini atoll. The ship was anchored about eighty-two nautical miles from the weapons test, just beyond the eastern boundary of the exclusion area. The exclusion area in this case, however, was set based on security considerations—to avoid having to evacuate the residents of Rongelap and Ailinginae—and not on potential fallout risks (Hewlett and Holl 1989, 171). The crew saw the flash and heard the detonation. About three hours later a fine white ash came down, coating the men, their gear, their boat, and the sea around them. Fearing that they might be detained by the Americans, or even that their ship might be sunk if their presence were known, the crew decided to head for their home port, Yaizu, near Tokyo, without informing anyone of what had happened. On 14 March they arrived, all suffering from radiation sickness, two so ill they went immediately to Tokyo University Hospital. Within a few days, all the others were in a hospital at Yaizu (Hewlett and Holl 1989, 175–77; Divine 1978, 4–11).

For the third time in a decade Japanese civilians had been exposed to the effects of atomic weapons and public response in Japan was emotional. The ABCC was drawn into the resulting turmoil, both by its official status as the American medical agency studying radiation effects in Japan and by its rela-

44. Scott Matsumoto, "Report," undated, and "Patient Rapport in a Foreign Country," drawer 12, Newspaper Clippings and Magazine Articles 1950–1957, NAS.

tionship to Tsuzuki, the Tokyo physician responsible for the treatment of the *Lucky Dragon* crew.

In early 1954, as head of the new Japanese Institute of Radiological Sciences, Tsuzuki published a report suggesting that fallout could cause radiation sickness.[45] Less than a month later, the "Yaizu incident"—the arrival of the twenty-three Japanese fishermen aboard the *Fukuryu Maru*—occurred. Their distinctive symptoms were quickly interpreted by Japanese physicians, particularly since the U.S. Atomic Energy Commission had announced a few days earlier that 236 Marshall Islanders were exposed to radiation from a Pacific weapons test. The news that a Japanese fishing crew had also been exposed made headlines in Japan and around the world. The fate of the *Lucky Dragon* crew provoked diplomatic maneuvers in both Japan and the United States and a consumer panic in Japan over radioactive tuna.[46] President Eisenhower suggested that radioactive fallout from the Bravo test had "surprised and astonished the scientists" (a statement played up in the American press). And many Japanese citizens expressed profound anger that Japanese had again been victimized by atomic weapons.

Tsuzuki took over care of the hospitalized fishermen and carefully limited access to them. The ABCC's director, then John S. Morton, and two physicians arrived in Tokyo on 18 March to examine the fishermen but were denied the right to take blood samples (Hewlett and Holl 1989, 175).[47] Morton unwisely

45. Tsuzuki mentioned the publication when he was interviewed by ABCC Labor Union Chair Nobutaka Ushio 13 August 1954. ABCC Director Holmes instigated the interview after a speaker at the 6 August peace rally referred to Tsuzuki as his source for some comments critical of the ABCC. The speaker said Tsuzuki had called the ABCC "an institute established for the purpose of researching the destructive power of the atom bomb." Tsuzuki said since he had never met the speaker, "I could have never said such a thing to him." He also told the ABCC interviewer that he had published a book entitled *Disasters of the Atomic Bomb from the Medical Point of View* on 5 February. In that book, he said, he argued that fallout could cause radiation sickness. The *Lucky Dragon* incident just a few weeks later, he said, had shown that he was right. In August 1959 Tsuzuki told a reporter his book was an "analysis and tabulation of the studies on the A-bomb disaster in Hiroshima and Nagasaki from the standpoint of medicine." I have not been able to find an English-language edition of this book, however. The 1954 Nobutaka interview is in the form of a memo to Maki dated 17 August 1954. The *Yomiuri Press* report is dated 16 August 1959, entitled "Interview with Masao Tsuzuki by Dr. Ichiro Nakayama." Both are filed in "Dr. Masao Tsuzuki," drawer 2, NAS.

46. Some American observers questioned the methods used to test for radioactivity and suggested that the fish were not in fact contaminated (Schull 1990, 134; Divine 1978, 11).

47. Also, "Report of Dr. Tsuzuki's Confirmation," translation, 13 August 1954, drawer 2, Dr. Masao Tsuzuki, NAS. This is an account of a meeting between Tsuzuki and Nobutaka Ushio, the chairman of the ABCC Labor Union, in which Tsuzuki explained comments attributed to him in a lecture delivered by Dr. Osada at the ceremonies 6 August in Hiroshima.

offered to provide medical care to the fishermen at the ABCC facilities in Hiroshima (see chapter 7), but his offer was refused. Tsuzuki respected Morton but found the physicians accompanying him "disagreeable." He believed the two young doctors underestimated his abilities as a scientist.[48] In a later newspaper interview, Tsuzuki said he proposed that the American physicians could be allowed to inspect the Japanese fishermen if Japanese physicians were allowed to inspect the Marshall Islanders suffering from radiation sickness as a result of the Bravo blast.[49] Such a request was consistent with Tsuzuki's basic stance that American scientists could have no special claims to study Japanese victims.

For their part, the Americans were concerned about what the remaining traces of radioactive ash on the *Lucky Dragon* might reveal about the design of the device that was tested. The AEC was especially sensitive about any evidence that might suggest the use of a new type of weapon. It refused to provide information about the weapon or the fallout to the Japanese physicians and scientists. This angered the Japanese, who suspected that a new type of bomb had been used and felt that if they knew more about the weapon they might be able to better help the injured crew (Hewlett and Holl 1989, 176). When Merril Eisenbud, director of the AEC's health and safety laboratory in New York, arrived in Tokyo 21 March, relations had degenerated so much that he was not even allowed to see the fishermen. The Americans were frustrated, both because they felt the crew might not be receiving adequate care and because they wanted to assess the health status of persons who had lived for two weeks in a highly radioactive environment. At the same time, Tsuzuki and his colleagues did not want to provide "guinea pigs" for another American experiment (Hewlett and Holl 1989, 176).

In August 1954, as several of the *Fukuryu Maru* crew still lay hospitalized, Tsuzuki tracked down ABCC director Robert Holmes (who had replaced Morton in July) while Holmes was in Tokyo on other business. Tsuzuki came to Holmes's hotel and personally asked him to visit one of the fishermen who seemed particularly ill. Holmes "recognized the [political] hazard of such a visit" but felt he could not refuse. He stayed briefly in the patient's room and offered his help "in an unofficial capacity." Tsuzuki asked if Holmes could help him acquire aureomycin, an antibiotic which he felt might help the pa-

48. Ushio Nobutaka to Maki Hiroshi, "Confirmation by Dr. Tsuzuki," 17 August 1954, drawer 2, Dr. Masao Tsuzuki, NAS.

49. "Interview of Dr. Masao Tsuzuki by Dr. Ichiro Nakayama," official translation of article in the *Yomiuri Press*, 16 August 1959, drawer 2, Dr. Masao Tsuzuki, NAS.

tient, and this seemed to be the end of the visit.[50] But as Holmes stepped out of the hospital, he was besieged by Japanese reporters who asked him whether the United States planned to provide monetary compensation to the victims and their families. Holmes said he was able to evade these questions by "registering a somewhat shocked appearance that they would be talking about so worldly a thing as money, when the critical issue was actually care of the patient."[51]

Only one of the *Lucky Dragon* crew died while hospitalized after the Bravo incident. This man, thirty-nine-year-old radio operator Aikichu Kobayama, may have died of infectious hepatitis rather than as a direct result of radiation exposure (Hewlett and Holl 1989, 271; Schull 1990, 133). Regardless of the precise cause of his death, however, he was a victim of Bravo, and the incident and its sequelae had a broad international impact. The twenty-three fishermen aboard the *Lucky Dragon* became symbols of further Japanese victimization and American inhumanity.[52]

A NEW NAME?

Meanwhile, another component of the organization's public image—its name—was being reconsidered. By 1955, virtually all concerned agreed that the name Atomic Bomb Casualty Commission was inappropriate. Both Japanese critics and AEC officials wanted to remove the words "atomic bomb" from the organization's title, though the two groups had different reasons for finding the term problematic.

The original selection of the name seems to have been relatively arbitrary, in the sense that there is no record indicating that a great deal of thought went into it. Lewis Weed, chairman of the National Research Council, proposed in June 1946 that President Truman create a federal "Commission on Atomic Bomb Casualties."[53] But this may not have been the source for the name finally

---

50. Aureomycin had alleviated some symptoms of radiation sickness, including diarrhea, in dogs experimentally irradiated in Atomic Energy Commission studies. ACBM, "Minutes," 20 October 1949, 59, box 3218, RG 326, ACBM, DOE.

51. Robert Holmes to R. Keith Cannan, 9 September 1945, drawer 3, NAS-ABCC Office 1954, NAS.

52. ABCC geneticist William J. Schull, in his memoirs, has recalled the *Lucky Dragon* incident as a sensational overreaction fueled by a Japanese press indifferent to the truth (Schull 1990, 131–33).

53. Lewis Weed to Norman T. Kirk, 28 June 1946, drawer 29, NAS.

adopted. Austin Brues, the University of Chicago medical professor (and director of the biology division at Argonne National Laboratory) who cochaired the investigative team sent to Japan in November 1946, told a colleague that the name was concocted as he and Paul Henshaw, of the U.S. Public Health Service, were "standing on a windy corner [in Washington], waiting for transportation." Brues said that Henshaw suggested "Atomic Bomb Casualty Commission" because "it would be called ABCC."[54] Neel recalls things differently, suggesting that the name was not proposed until later, when he, Brues, and Henshaw were in Japan.[55] Henshaw shares Neel's recollection, more or less, writing in a 1991 memoir that he thought up the name while in Japan in November 1946 (Henshaw 1991, 12).

Whatever its origins, the name Atomic Bomb Casualty Commission was "not a formal" name, but simply descriptive of the program to be carried out by the NRC in Japan.[56] But by 1953, when the AEC launched its "Atoms for Peace" campaign, many within the AEC felt that the words "atomic bomb" should be removed from the title because they emphasized the project's military implications rather than its importance to the peaceful development of atomic energy. John Bugher particularly wanted a new name and felt the change should be timed to coincide with the implementation of AEC plans to make the ABCC "a base for the distribution of isotopes [for medical use] in Japan and perhaps to other adjacent countries of the Far East."[57] Keith Cannan of the NRC also felt the name should emphasize the peaceful uses of atomic energy, "atoms for peace," instead of the "atomic bomb."[58]

Japanese advisers to the ABCC began to complain about the name as well. Tsuzuki objected to any name with either "A-bomb" or "radiation" in the title. And Nagasaki Medical School dean Kitamura Seiichi said the study would

54. In an anecdotal speech he delivered to the Chicago Library Club in 1961, "The Chrysanthemum and the Feather Merchant." This speech is mentioned in a 10 March 1969 lecture by Charles Dunham, who was director of the AEC Division of Biology and Medicine in the late 1950s. Atomic Bomb Casualty Commission, "ABCC Meeting, 3–13–69, Washington, D.C.," box 33, Thomas Francis papers, BHL.

55. James V. Neel, interview by author, 22 November 1988, Ann Arbor, Michigan.

56. Bugher suggested that the name was chosen so that it would be different from the name "Joint Commission," that is, the group that conducted the original studies in Japan. See Japanese Advisory Council, "Minutes, Second Meeting," 20 February 1956, 22, Japanese Advisory Council Minutes 1–10, NAS.

57. The AEC was also promoting the creation of a "school" for the medical use of radioactive isotopes, to be set up at the ABCC in Hiroshima, and the construction of a reactor in Hiroshima for medical use. "Minutes, 49th Meeting," 11–12 March 1955, ACBM, DOE.

58. Keith Cannan, "Memorandum on My Visit to ABCC, October 15 to November 10, 1955," drawer 28, Cannan-Hastings Report, NAS.

be more successful if it were conducted "under a name other than ABCC."[59] "A-bomb" did not seem to be a public relations problem for the Japanese-run A-Bomb Hospital or the Japanese Council for A-Bomb Survivors, as another Japanese physician pointed out. It was the use of the words by an American agency that did not provide medical treatment to the survivors and other research subjects that was perceived by Japanese critics of the ABCC as a problem.[60]

In February 1956 the question of a name change appeared on the agenda of the ABCC's Japanese Advisory Council, a group of leading Japanese scientists and physicians who met annually to discuss the ABCC program. Those attending agreed that the name should be changed. Morito Tatsuo, president of Hiroshima University, suggested "Radiation Medicine Research Institute." Tsuzuki proposed "Nuclear Medical Institute" or "Atomic Medical Institute." Thomas Parran, then dean of the School of Public Health at the University of Pittsburgh and attending the meeting at the AEC's request, said that "atomic" had a "bad connotation" and should be replaced by "nuclear" or "radiation." University of Tokyo medical professor Nakaidzumi Masanori proposed "Institute for Radiation Injuries," and Hiroshima University Medical School dean Kawaishi Kunio countered with "Atomic Biomedical Institute." Shields Warren, also attending the meeting for the AEC, suggested "Institute for Human Radiobiology." Others names proposed included "American-Japanese Medical Institute" and "Japanese-American Medical Radiobiology Institute." Those attending were asked to choose the name they favored and inform the council chairman, Kojima Saburo, director of the Japanese NIH, within two weeks.[61]

Some six weeks later, in April 1956, Shields Warren, as chairman of the NRC's Committee on Atomic Casualties, recommended to the AEC that one of the new names proposed by the Japanese Advisory Council be accepted. Warren's own choice was "Japanese-American Institute for Medical Radiobiology." "This program at Hiroshima and Nagasaki has long since left behind its military significance," Warren said, and the bomb should be removed from the title to reflect this change.[62]

Somehow these suggestions did not have much impact, for the question of

59. Japanese Advisory Council, "Minutes, Second Meeting," 20 February 1956, 9, Japanese Advisory Council Minutes 1–10, NAS.

60. Masaoka Akira and Kojima Saburo, in Japanese Advisory Council, "Minutes, Third Meeting," 14 November 1956, Imperial Hotel, Tokyo, file 106, ABCC 1957, box 9, JCB.

61. See Japanese Advisory Council, "Minutes, Second Meeting," 20 February 1956, Japanese Advisory Council Minutes 1–10, NAS.

62. John C. Bugher, "Report: Visit to Atomic Bomb Casualty Commission Operations, Hiroshima and Nagasaki, Japan, February 1956," 9 August 1956, 200, file 107, box 16, General Corre-

a new name came up again at the next meeting of the JAC.[63] This time ABCC director Robert Holmes suggested that the name should remain the same. The ABCC staff and its advisers had considered a name change "for over a year," and at one time all parties agreed that "A-bomb" should come out of the title. However, Holmes said, "As time has gone by the need to change the name seems to have become less and less important. Our view now is that there is no pressure for a change." The ABCC had "developed during the past years a reputation for itself internationally," and people were "used to referring to it as ABCC." It might therefore be "difficult to change in the minds of other nationalities" (19). The Japanese advisers responded sympathetically. "I think a new name would be better," observed Masaoka, but "we must . . . weigh this question carefully because, as Dr. Holmes has mentioned, the ABCC has built up some reputation internationally" (20). The president of the Japan Science Council, Kaya Seishi, then pointed out that a strictly cosmetic name change could have negative repercussions: "I believe that we should not consider a change in name at this time unless there is a great major change in policy, operation and program." A "superficial" name change, he said, "would have a strange [effect]." Tsuzuki held firm and continued to support a name change, but when the chair suggested that the discussion end there were no objections. The name change remained technically as a matter pending further review, but it did not come up again in meetings of the JAC.[64]

John Beatty has suggested that the name was not changed in the 1950s because the ABCC staff and American diplomatic staff in Japan believed "that the Japanese would not fail to recognize the change as an essentially political move" (1991, 292). I would agree that the Japanese, as Tsuzuki's response indicates, would immediately have recognized that changing the name was a political act. This does not, however, entirely explain the lack of action. Rather, I suspect that both sides recognized that a name change would be a statement about the future of the ABCC, and the impasse was a consequence of conflicting ideas about what the ABCC was and what it should become. The organization remained the Atomic Bomb Casualty Commission until 1975, when it was renamed the Radiation Effects Research Foundation. This eventual change reflected a significant rearrangement of the organization and funding of the program, but as Beatty has demonstrated, it was also charged with political

spondence, JCB. This was Bugher's report on the trip with Shields Warren, written for the Rockefeller Foundation.

63. Masaoka Akira and Kojima Saburo, in Japanese Advisory Council, "Minutes, Third Meeting," 14 November 1956, Imperial Hotel, Tokyo, file 106, ABCC 1957, box 9, JCB.

64. Japanese Advisory Council "Minutes, Third Meeting," 14 November 1956, Japanese Advisory Council Minutes 1–10, NAS.

meaning: the new RERF celebrated the notion of the ABCC as a special form of international science, an internationalism demonstrated in its funding, organizational structure, and new bilingual letterhead (Beatty 1993).

A name change that was primarily cosmetic in 1956 or 1957 could have had advantages for both the American sponsors of the ABCC and its Japanese advisers and participants. The problem was that the advantages those groups sought were not congruent. Tsuzuki was interested in a name that would suggest that the ABCC was a provider of medical care, even thought it was not; several American officials were interested in a name that would reflect their new emphasis on "atoms for peace," even though the ABCC's research had military implications. None of the proposed changes would necessarily have affected the day-to-day operations of the ABCC in Japan. But each would have defined the ABCC and its mission in different terms. The fundamental question underlying the debate over the ABCC's name was the question of mission, and on the ABCC's proper mission American administrators and Japanese scientists often disagreed.

CONCLUSION

ABCC staff members sometimes responded to public criticism of the organization's policies by stressing the "strictly scientific" nature of the enterprise or by renaming procedures or departments. But the "pure science" defense angered many survivors and allowed critics to exploit negative perceptions of scientists in public attacks on the organization, and calling procedures by different names did not resolve the basic problems in ABCC "patient relations." Widespread objections to the autopsies were not eased by calling them "examinations,"[65] nor was anger over the lack of treatment eased by calling diagnosis "treatment."

This reliance on verbal adjustments was particularly problematic given the bilingual nature of the enterprise. As investigators discovered, words or symbols that were neutral in English might be offensive in their Japanese form. The use of the word "biostatistics" as the name of one department, for example, offended many Japanese, who apparently thought this word suggested science could reduce human beings to biological statistics. After several complaints,

---

65. This policy is discussed in "Instructions to Genetics Doctors Regarding Reporting of Autopsy Findings," 27 September 1951, 5/VB, JVN: "In all reports to *parents* the word 'autopsy' is to be avoided and 'examination' used instead" (emphasis in original).

the ABCC changed the name of the department to "statistics."[66] The ABCC also stopped using the code number four during anthropometric examinations, after American staff learned that the number four, or the showing of four fingers, was a way of referring to the Eta or burakumin class, a despised group.[67]

In the mid-1960s, charges that the ABCC was hoarding data about the biological effects of radiation became the basis of a "white paper" campaign. It was a period when several such investigations had uncovered government scandals in Japan. ABCC critics were particularly interested in seeing the ABCC's raw data from the Occupation period, including the results of the 1950 census, which attempted to determine how many survivors there were. The campaign was aided by various factions of the Communist Party in Japan, and much of the communist criticism was directed not at the ABCC, which was dismissed as a worthless American organization, but at the Japanese government, which had failed to carry out the medical studies itself. It was, one communist official said, a "national disgrace" that the government left the study to the United States. A communist handbill passed out at Hiroshima railway stations in October 1967 urged the government to "revoke the unpatriotic, inhumane agreement" whereby the "tissues from dead bodies" were taken by the ABCC.[68] The themes in this campaign—of concealment, stolen body parts, and national humiliation—were in many ways products of historical circumstances over which the ABCC itself had little control; the ABCC's relation to the communities of Hiroshima and Nagasaki was deeply affected by the broader problems of Japanese-American relations.

At the same time, the survivors' sense of victimization in Hiroshima was not eased by many ABCC practices. The organization overlooked Japanese needs in small details. The flooring in the waiting room for mothers and babies, for example, was highly polished linoleum. Japanese women in their wooden clogs with metal taps sometimes slipped and fell. Waiting room magazines and signs were in English. More problematically, the ABCC did not treat the survivors it studied; it summoned patients to the clinic for examination during

66. "Is ABCC broad-minded enough to reconsider the extent of misunderstanding which the world 'bio' has created?" asked a reporter for the *Chugoku Press* in August 1959. "The Actual State of ABCC," translation of article in *Chugoku Press*, 7 August 1959, drawer 2, ABCC Press mid-1957–1964, NAS.

67. Prior to the Meiji era (1867) people in certain occupations such as the slaughtering of animals or dealing in animal hides were outcasts in Japanese society. Considered unclean by Buddhist tenets, they were known as the "Eta." Government ordinance outlawed overt discrimination in 1871, but the Eta remained in segregated neighborhoods and rarely intermarried with other classes. Scott Matsumoto discusses this problem in "Patient Rapport in a Foreign Country," drawer 12, Newspaper Clippings and Magazine Articles 1950–1957, NAS.

68. Translation in drawer 12, Japanese Press Clippings 1967, NAS.

weekday working hours, which often meant the loss of a day's pay; and it offered patients little compensation other than the "knowledge that they are participating in a scientific endeavor" (Matsumoto 1954, 69).

The ABCC did, however, provide its subjects with valuable medical information. The nine-months examination of Japanese infants amounted to a free well-baby clinic, at which an American physician assessed an infant's development and general health. For many babies in Hiroshima and Nagasaki, this was the first medical examination since the at-home ABCC visit shortly after birth. Adult participants also benefited, from thorough and frequent medical examinations.

The ABCC staff did try to improve its image in the popular press and among those participating in the study. The staff reached out to the community, sometimes awkwardly, to engage its support. Yet fundamental ideological and cultural differences, most profoundly expressed in the ABCC emphasis on "pure science," separated the American staff of the ABCC from the survivors and their communities. That science itself is the subject of my next three chapters.

The most visible science of the ABCC was its genetics project, a project shaped by the Cold War and by public concerns about radiation effects. By examining the definition in practice of mutation, the first analysis of the genetics data, and the publication strategies that guided the two primary texts, I demonstrate that the cultural meaning of radiation, as a terrifying product of atomic weapons, was written into the language of the science. The atomic bomb was a part of the scientific analysis, wrapped into the logic and rationality of the debates, and present in the public texts that transformed the events at Hiroshima and Nagasaki into data that could be applied to medical care, civil defense planning, and worker protection legislation around the world.

 THREE

Science

and

Context

CHAPTER NINE

What is a Mutation?

Mutations have been important to the science of genetics, to the construction of evolutionary theory, and to efforts to understand the biological effects of radiation. Scientists first localized the abstract gene by studying how certain mutant effects were inherited. While a trait common to an entire population could not be analyzed, a visually discernible deviation could be.[1] At the same time, mutation has been a remarkably plastic concept, interpreted differently depending on the problem being investigated, the organism of interest, and the consequences of the interpretation.

This chapter examines the working concept of mutation in the morpholog-

---

1. As Richard Goldschmidt observed at the 1951 Cold Spring Harbor Symposium on Quantitative Biology, mutations provided evidence for the existence of genes. "All our knowledge of the genetic material . . . is derived form the study of mutation. In classical genetics the gene is an

ical genetics study of the ABCC. The signs chosen in the ABCC genetics study suggest that Neel, Schull, and their collaborators and advisers were using mutation to measure not simply genetic change, but also the potential social and evolutionary costs of the exposure of human populations to radiation. While Neel and Schull have tended to see these two goals as entirely congruent, I suggest here that the goal taken to be primary had an impact on their assessment of the data collected. They recognized as mutations disadvantageous departures from the norm that seemed to have implications for either immediate health or the long-term genetic health of the human species. This was an approach supported by their scientific advisers in the United States.[2] Indeed, it was an approach shared by other researchers studying mutation in human populations exposed to radiation.[3] Yet it was not entirely consistent with animal studies, most notably those of the *Drosophila* and mouse geneticists.

My interest here is not in the relative scientific legitimacy of these approaches, but in how they illuminate the profound ambiguity of the concept of mutation. Mutation was not, in these studies of human populations, *any* heritable change in the genetic material. In committee meetings and correspondence those involved continued to invoke this formal definition of mutation, but how scientists use a term in constructing and selecting data may be more revealing than how they define it formally. In day-to-day practice, in data analysis and decision making, a mutation was instead a dangerous morphological trait with implications for the future survival of the species, a specifically threatening change. This implicit definition of mutation as socially dangerous had practical consequences. It shaped what bodily signs were taken as evidence of hereditary change.

GENES AND MUTATIONS

Access to the human genome in 1946 depended on the recognition of morphological variations—changes in the body—that might be called mutations and linked to an unseen gene for the purposes of genetic analysis. Neel's original proposal for the genetics project, therefore, identified morphological traits as

---

extrapolation from the mutant locus" (Goldschmidt 1952, 1). Theodosius Dobzhansky makes a similar point, though far less succinctly, in his *Genetics and the Origin of Species* (1937, 19–49).

2. National Research Council, Division of Medical Science, "Records of the Conference on Genetics," 24 June 1947, Genetics—1947 #1, Drawer 19, NAS. See also National Research Council (1947).

3. Most particularly the survey of Stanley Macht and Philip Lawrence, and of James F. Crow (Macht and Lawrence 1955; Crow 1955).

mutations, though this was clearly understood by Neel and his advisers to be a form of shorthand, since the mutation was something that occurred at the invisible level of the gene. In a presentation to scientific colleagues in June 1947, Neel suggested the ABCC genetic project collect data on sex ratio and rates of stillbirth, congenital anomaly, and infant mortality.[4] Six days after this important presentation, Neel prepared a final report that included, in response to the discussion at the genetics conference, "bodily dimensions and life span," which were considered "fitness indicators." While dominant mutations with gross phenotypic effects would be easier to observe, Neel said in this second report, they could be expected to be much rarer than those with "smaller, but still possibly serious, effects on bodily dimension, life span, etc." The ABCC should track birth weight and growth and development at nine months, since mutations affecting such fitness parameters could be expected to appear in living populations more frequently than mutations producing dramatic phenotypic anomalies.[5]

Each of the indicators selected for analysis in the ABCC genetics study, then, had some theoretical justification. They were not all equally practical, nor were they all equally representative of the results with model organisms. Statistical changes to be expected in the offspring of survivors would be higher rates of major malformation, stillbirth, and neonatal death; lower birth weight; retarded growth at nine months; and changes in sex ratio as follows: for irradiated mothers a "reduced" sex ratio (fewer male offspring than normal), for irradiated fathers an "increased" sex ratio (more male offspring than normal).

These indicators would presumably reflect several kinds of mutational events. An increased rate of major malformation, stillbirth, or neonatal death could reflect newly induced dominant mutations anywhere on the genome. Sex-ratio changes might reflect the induction of recessive lethal mutations on the maternal X chromosome. The "fitness" indicators of birth weight and growth at nine months could reflect slightly deleterious, but not disastrous, mutations anywhere on the genome.

Schull felt that they were "almost certain to want to weight some indicators more heavily than others since they are not all equally efficient in re-

---

4. "Stillbirth" was used in the Japanese sense of a "deadborn foetus of any age after the third month" of pregnancy, thus including spontaneous abortions after the first trimester. All these initial parameters are explored in more detail in James V. Neel, "Report, 10 June 1947, Concerning the Study of the Genetic Effects of the Atomic Bombs in Hiroshima and Nagasaki," drawer 5, JVN.

5. This report was apparently written by Neel but signed by the entire committee: George W. Beadle, Herman J. Muller, C. H. Danforth, L. H. Snyder, Donald R. Charles, and Neel. "Concerning the Study of the Genetic Effects of the Atomic Bombs in Hiroshima and Nagasaki," 30 June 1947, drawer 19, Genetics—1947 #1, NAS.

flecting radiation damage," but acknowledged that "the choice of the weight is difficult." Any "purely arbitrary" selection would be "biologically suspect," yet how important each indicator should be in their relative assessment of radiation damage was unclear.[6]

The sex ratio was perhaps the strongest indicator—both theoretically and practically. As a practical matter, sex is rarely misdiagnosed. The theoretical justifications were equally compelling. Though Donald Charles was unable to detect sex-ratio effects in his wartime work with mouse irradiation, Muller had amply documented sex-ratio effects in *Drosophila* (Muller 1927; Charles et al. 1960). In human populations the effect was expected to work the same way it worked in *Drosophila*. Human males, like *Drosophila* males, have an X chromosome from the female parent and a Y chromosome from the male parent. The Y chromosome was believed to carry relatively little genetic information—at least, the degree of genetic homology with X was unknown. All traits carried on the X chromosome derived from the mother were therefore expected to be expressed, in the absence of a compensatory X chromosome to inhibit the expression of recessive traits. New, lethal recessive mutations induced on the X chromosomes of heavily irradiated mothers would presumably lead to a greater prevalence of early embryonic death of male fetuses. This would result in proportionately fewer male birth to heavily irradiated mothers. Since most mutations were presumably both recessive and inconsistent with survival, sex ratio in the offspring of irradiated mothers might be a particularly sensitive index of mutation rate.

Neel also postulated that the sex ratio in the offspring of heavily irradiated fathers married to nonirradiated mothers would change. With fathers he expected the change to be in the opposite direction. A human female has two X chromosomes, one from each parent. Any undesirable recessive traits on these chromosomes were expected to be suppressed by the corresponding trait on the matching X chromosome. But newly induced dominant mutations on the X chromosome contributed to daughters by fathers would presumably result in a greater prevalence of early embryonic death of female fetuses. Irradiated fathers should therefore have slightly fewer daughters. This sex-ratio parameter would be less sensitive, however, since it depended on the induction of dominant lethal mutations.[7]

6. William J. Schull to Duncan McDonald, 17 June 1952, 6/Q, JVN.

7. The findings by British geneticist Mary Lyon confuse this picture. "Lyonization" refers to the process by which one X chromosome in mammalian females is inactivated in the course of embryonic development. Whether the X chromosome is that contributed by the mother or father seems to be random, and since inactivation occurs at different stages in different cell lines, all mammalian females are genetic mosaics, that is, different cells manifest the traits of either the

The weakest indicators, practically and theoretically, were the fitness indicators of birth weight and growth at nine months. Fitness itself was an ill-defined biological term, and these two indicators were recognized to be of complex etiology, influenced by both environment and heredity. In addition, the measurements posed certain practical problems. At first, midwives were using nonidentical, uncalibrated scales to weigh the newborns; gradually these were replaced by standard scales supplied by the ABCC.[8] But some midwives did not write down the weights at the time of the birth.[9] Some who did reported weight in grams, others in a Japanese unit, the momme.[10] And there was no generally agreed upon time for the weighing to take place. Newborns lose weight rapidly in the first twenty-four hours of life, so the weight figures, even if accurate, were not necessarily comparable. In addition, some midwives were not recording the length of the newborns.[11]

The anthropometric measurements were also problematic. Japanese pediatricians were told to measure newborns' head circumference, chest circumference, and length on the home visits, Neel said, in order to "insure that [they] looked at the baby a little more closely than might otherwise be the case, and also as an aid in the early diagnosis of hydrocephalus." But the measurements themselves were, he felt, inaccurate—"for the most part not worth the paper on which they are written"—and they were never used in the analysis.[12] Some children were measured again, in the ABCC follow-up clinic, at nine months,

---

maternal or paternal X chromosome. What this might mean in terms of effects on the sex ratio in the offspring of irradiated females is unclear (Lyon 1961).

8. I have been unable to find documentation indicating precisely when these scales were supplied, but they were available to the midwives at least by 1950 and possibly before. See "Report on a Meeting of Midwives Association at Hijiyama," 22 September 1951, drawer 5, JVN. The midwives were allowed to keep the scales when the ABCC terminated data collection on newborns in 1954; Duncan McDonald to Newton Morton, Grant Taylor, and James V. Neel, "Midwives Meeting at Hijiyama, 21 November 1953," 2 December 1953, drawer 19, Genetics—1949–1953 #3a, NAS.

9. "Birthweight is approximate and cannot be judged accurate. Thus, one midwife returned a batch of registration questionnaires to the City Hall registration office, and when it was pointed out to her that she had omitted the birth weights, she promptly pulled figures from out of her head and wrote them down." See "Genetics Follow-Up Program," 2 September 1948, drawer 6, JVN.

10. Ray Anderson, "Remarks for Midwives," 21 May 1951, notebook #1, Genetics Program, WJS, AA.

11. Masuo Kodani, "Memorandum for the Record," 22 March 1949, drawer 5, JVN.

12. James V. Neel and William J. Schull to Newton Morton and Duncan McDonald, 30 August 1952, JVN. Newton Morton was at the time a graduate student who took a year off to work in Hiroshima. He returned to the United States to study with James Crow. Duncan McDonald was a Canadian biologist who was directing the ABCC genetics program in Japan while Neel and Schull worked on data analysis in Michigan.

and these measurements Neel deemed more accurate. Two nurses, one specialized in anthropometric measurement, were responsible for weighing the infants and measuring body length, chest circumference, and head circumference.[13] But the follow-up program did not begin until the spring of 1950, and originally it involved the examination of one-year-olds.[14] Because younger babies were easier to handle, the examination date was moved up to nine months. As a result, the ABCC nine-months anthropometric data covered less than two years of the total data collection period. To complicate matters, infants did not necessarily appear at the ABCC clinic when they were exactly nine months old. Due to predictable complications in notification and scheduling, those examined ranged in age from eight to ten months at the time of examination (Neel and Schull 1956a, 164). In rapidly growing infants, a few weeks could make a significant difference in size. Neel and Schull tried to compensate for this difference in their analysis by dividing the infants by age at examination (7.5 months to 8.5 months; 8.5 to 9.5; and 9.5 to 10.5), as well as by sex, exposure history of parents, and city of examination (Neel and Schull 1956a, 179).

Both birth weight and growth at nine months posed serious practical problems, but they were retained in the ABCC study because they were supposed to provide access to a type of mutational event assessed as extremely important: minor deleterious mutations having some cumulative effect on fitness. From the viewpoint of the individual affected, such mutations were less problematic than either lethal mutations or dominant mutations causing gross anomaly. But from an evolutionary or populational perspective, they could be interpreted as more threatening, since they were consistent with reproductive success.

Major malformation, stillbirth, and neonatal death, the remaining indicators, presented varying problems of both ascertainment and etiology. They are interesting partly for their relationship to three indicators not counted as signs of genetic change at Hiroshima and Nagasaki—minor malformation, reduced fertility, and early spontaneous abortion—all of which were taken into account in planning and managing the study but rejected in the eventual published analysis. A fourth possible indicator for measuring mutation rates, "sentinel phenotypes," was never considered a serious possibility by Neel and Schull. Like the parameters selected for analysis, these four rejected indicators posed various problems in both theory and practice.

13. A more detailed description of this process is included in W. W. Sutow to Duncan McDonald, 9 December 1952, notebook #9, Genetics Long Form, WJS, AA.

14. William J. Schull, "Memorandum for the Record," 7 March 1950, notebook #9, Genetics Long Form, WJS, AA.

SENTINEL PHENOTYPES

Of the four the exclusive use of sentinel phenotypes was most clearly impracti-cal. Sentinel phenotypes are specific anomalies believed to result from single, dominant mutations for which geneticists have estimated spontaneous muta-tion rates in large human populations. By 1954, when the first serious analysis of the ABCC genetics data was underway, these included such conditions as achondropastic dwarfism; chondrodystrophic dwarfism; two malformations of the eye, retinoblastoma and aniridia; and neurofibromatosis.[15] Presumably, the ABCC geneticists might have compared the rates of appearance of these senti-nel phenotypes in the survivor populations to the known spontaneous mutation rates in other populations. If the sentinel phenotypes appeared more frequently in the offspring of the exposed this might be compelling evidence of radia-tion effects.

Neel recognized early that a crude comparison of the total number of ab-normalities was less powerful than a comparison of specific anomalies. "More and more I come to feel that a comparison of figures of the overall incidence of gross malformations is very misleading," he told Schull in 1949. "The really valid comparison concerns the incidence of such specific malformations as harelip, anencephaly, club foot, and so on."[16] Yet the use of sentinel or specific phenotypes alone as a sign of mutation was impractical in the ABCC study because the numbers of births exhibiting these phenotypes among the exposed and control populations at Hiroshima and Nagasaki were too small to provide a meaningful comparison. Earlier studies that had successfully estimated spon-taneous rates of mutation for sentinel phenotypes generally involved extremely large populations. The data for chondrodystrophic dwarfism, for example, were drawn from records of 94,075 births in Copenhagen. Ten of these chil-dren were chondrodystrophic dwarfs, and eight of these were born to parents who were not affected, indicating that about one birth in twelve thousand might result in a mutant chondrodystrophic dwarf (Mørch 1941). The ABCC tracked about seventy-six thousand births altogether—presumably a large enough group for some significant overall rate to emerge—but these births were divided according to five categories of parental radiation exposure and two cites, in which the radiation mixes differed.[17] Despite the relatively large total population studied at Hiroshima and Nagasaki, sentinel phenotypes were

15. Neel was himself responsible for estimates of spontaneous mutation rates for retinoblas-toma (Neel and Falls 1951).

16. James V. Neel to William J. Schull, 31 August 1949, 6/K, JVN.

17. Because the two bombs were not identical in structure, survivors exposed at Hiroshima received more gamma radiation than those exposed at Nagasaki. John Auxier describes the labori-

not frequent enough in any particular group to provide a basis for comparison (Strobel and Vogel 1958).

Because the sentinel phenotypes were important and well-known genetic anomalies, generally recognizable, the ABCC investigators collected information about their occurrence in the normal course of data collection. But in the first four years of the study, from 1948 to 1952, the ABCC diagnosed only thirteen cases of achondroplastic dwarfism in their entire sample, including the offspring of those exposed at various distances and of all controls.[18] By July 1953, one group of children examined at the ABCC facilities at Hiroshima included only two chondrodystrophic dwarfs, two cases of micropthalmus (abnormally small eyes), and not a single case of retinoblastoma—all traits for which spontaneous mutation rates had been estimated.[19] While sentinel phenotypes had to be counted and included in the data, they could not be used independently as the basis of analysis.

STERILITY

Neel's decision to interpret reduced fertility or sterility in survivors as only tangentially relevant to the genetics project is somewhat more puzzling. Both Charles and Muller had amply documented genetic effects of radiation on fertility. Though early sterility after heavy radiation exposure is due to an absence of spermatozoa (reflecting the cessation of sperm production rather than the induction of mutations), reproductive failure after three or four months could reflect genetic mutations or chromosome breaks inconsistent with fetal survival and therefore be an important sign of the genetic effects of radiation. In 1950 C. D. Darlington postulated that radiation exposure would "always reduce fertility" and D. E. Lea in his 1947 summary of the status of radiation research noted that effects on fertility were well documented in both flies and plants.[20] Much of this effect was attributed to dramatic chromosomal rearrangements which made cell replication impossible. Darlington also ob-

---

ous "nuclear archeology" through which radiation mixes and dosimetry were estimated in the decades after the bombings in *Ichiban: Radiation Dosimetry for the Survivors of the Bombings of Hiroshima and Nagasaki* (1977). For an earlier and influential assessment of radiation dosimetry, see Robert R. Wilson, "Nuclear Radiation at Hiroshima and Nagasaki" (1956).

18. Duncan McDonald, "Cases of Achondroplasia, 1948–1952," 1 September 1953, WJS, AA.

19. "Report," July 1953, in Case Numbers of Major Congenital Anomalies on Long Form, WJS, AA.

20. C. D. Darlington, "The Cell and Heredity under Ionization," in Haddow, (1952, 43); see also Lea (1947, 328).

served that "bad genes are more easily passed on than bad chromosomes." When bad genes appeared in the same cells as bad chromosomes (as a result of radiation exposure), this would cause reduced fertility rather than malformed babies. This was both natural and desirable, Darlington suggested, for it would "reduce the deferred damage to the species from gene mutation."[21] Reduced fertility, then, was a natural purification of the gene pool and, as such, not a threat to human survival.

The ABCC did document fertility effects in atomic bomb survivors but the fertility studies were not a part of the genetics study.[22] Neel and Schull did not include any analysis of fertility rates in their 1956 monograph on the genetics study. In 1948 Neel said the proposed fertility study would "tie in very closely" with the genetic study but would not be a part of it.[23] One of Neel's objections to reduced fertility as a sign of mutation was that a drop in fertility might reflect an increase in induced abortions. The elective abortion rate in postwar Japan was particularly high after 1948, when the Japanese government eased restrictions on abortion (Neel and Schull 1956a; 19; Koya 1953). In addition, there were conjectures that exposed survivors might be more likely to avoid reproduction, because they might fear the genetic effects of their exposure on future offspring. While later studies revealed no differences in contraceptive practices among different exposure groups, this was an early concern in the fertility studies (Blot and Sawada 1972).

I would suggest that infertility in survivors was not a "genetic" problem because it did not involve the transmission of traits to the next generation and the addition of new mutations to the human gene pool. Certainly Neel expressed a related attitude toward early spontaneous abortion. Neel viewed early spontaneous abortion as "close to the genetic end of things," but the ABCC's single project to track early spontaneous abortion was placed under the obstetrics-gynecology department of the ABCC rather than the genetics department.[24] Neel's objections—particularly his sense that early spontaneous abortion was not as relevant to the genetics project as other indicators—reflected the tendency of geneticists to equate mutation in human populations with pollution of the gene pool and with human suffering. Both infertility and early spontaneous abortion, therefore, could be interpreted as less grievous

21. Darlington, "The Cell and Heredity under Ionization," in Haddow (1952).

22. See for example D. G. Seigel, "Frequency of Live Births among Survivors of Hiroshima and Nagasaki Atomic Bombs" (1966); W. J. Blot and H. Sawada "Fertility among Female Survivors of the Atomic Bombs at Hiroshima and Nagasaki" (1972).

23. James V. Neel to Masuo Kodani, 20 August 1948, drawer 19, Genetics—1948 #2, NAS.

24. Kitamura complained mildly about this classification in a letter to Neel, 25 February 1950, 5/QCC, JVN.

(for both the individual and the species) than, for example, malformation or neonatal death.

EARLY SPONTANEOUS ABORTION

Early spontaneous abortion was an obvious possible indicator of radiation-induced mutations in the hereditary material. Spontaneous abortion might be presumed to reflect the presence of new mutations incompatible with fetal survival. Collecting information about early abortion, however, is difficult in any population. Many such abortions occur before the woman knows she is pregnant, and most occur in the first twelve weeks of pregnancy. Tracking spontaneous abortions was particularly difficult in Hiroshima and Nagasaki, where most pregnancies were attended by midwives rather than physicians.

Neel saw the difficulties in Hiroshima and Nagasaki as virtually insurmountable when he planned the ABCC study. But a brief, moderately successful survey was completed by a Japanese physician, Kitamura Saburo, who was affiliated with the ABCC for about one year. The early spontaneous abortion study continued for another two years after Kitamura's departure, under the direction of two Japanese physicians employed by the ABCC. Kitamura's results differed in rather peculiar ways from the results collected by the later investigators. While the 1953 report to the genetics conference dismissed the spontaneous abortion data as indicating no difference in exposed and controls, the situation was actually much more complicated.

Kitamura was a Hiroshima physician, educated in the United States and married to an American. He had attended the University of Pennsylvania undergraduate and medical schools and worked as an intern at Boston Lying-In Hospital. In 1941, shortly before he planned to accept an academic appointment in the United States, he and his wife traveled to Japan to visit relatives. They were unable to return to the United States after the Japanese attack on Pearl Harbor in December 1941. During the war, Kitamura practiced as an obstetrician-gynecologist in Hiroshima, delivering some three thousand babies. But his status as a highly Americanized Japanese (having spent eighteen years in the United States) with an American wife resulted in considerable hardship.[25]

25. James V. Neel, "Memorandum for the Record," 6 November 1947, Conference with S. Kitamura, 6/C, JVN.

Kitamura made "quite a favorable impression" on Neel at their first meeting in the fall of 1947, and at Neel's suggestion Kitamura began planning a study of early spontaneous abortion in the summer of 1949.[26] He met with the Eugenics Committees of Hiroshima and Kure (at the time the ABCC was still maintaining operations in Kure as a control city), with positive results.[27] Over the next few weeks, Kitamura contacted Hiroshima's thirty-five practicing obstetricians-gynecologists and asked them to inform him of all spontaneous abortions that came to their attention. He did not attempt a similar survey in Kure.[28]

He visited each physician personally on a regular basis asking about spontaneous abortions, but these visits had to be handled carefully, since the doctors were "hypersensitive because of the income tax involved."[29] When physicians gave him the names of patients who had experienced early abortions, Kitamura visited the families at their homes, where he interviewed the wife about her exposure to radiation and her past medical history. In these initial visits, Kitamura did not interview husbands, nor did he ask wives to describe their spouses' exposure histories.

The numbers of early abortions reported to him increased steadily. In October 1949 he tracked twenty-six early abortions, and in February 1950 he collected information on about sixty-one spontaneous abortions.[30] The Eugenics Committee of Hiroshima Prefecture had estimated that sixty to seventy spontaneous abortions occurred each month in Hiroshima. Kitamura fairly quickly reached this level of reporting, and maintained or bettered it for the rest of his involvement in the program.[31] By the time he terminated his work with the ABCC, Kitamura's average reporting load had reached ninety-five cases per month.[32]

26. Ibid.

27. Kitamura Saburo, "Research Project Outline," 23 August 1949, Kitamura Program, 5/QCC, JVN.

28. Kitamura summarized his study plan in letter to Neel, 25 February 1950, Kitamura, 5/QCC, JVN.

29. Ibid.

30. Ibid.

31. Kitamura Saburo, "Research Project Outline," 23 August 1949, Kitamura Program, 5/QCC, JVN, contains the Eugenics Committee estimate. In February 1950 Kitamura collected information on eighty-one spontaneous abortions. See "Appendix G, Project OG-16," 20 March 1950, Kitamura Program, 5/QCC, JVN.

32. Duncan McDonald, "Early Termination Program: Investigation of Cases Not Submitted for Investigation, July 1–Dec. 31, 1951." 12 September 1952, Kitamura Program, 5/QCC, JVN.

After a year, in the fall of 1950, he broke off relations with the ABCC. Though there are numerous references to the fact that Kitamura "severed his relations" with the ABCC, I have not been able to find any documents explaining precisely why this occurred. Kitamura had collected data on 1,053 early spontaneous abortions, and the project seemed to be going well. His most important mistake—his failure to interview the husbands—had been noted, and the ABCC staff had devised a plan to rectify it.[33] Some of the Japanese physicians employed in the genetics program went back to the homes Kitamura had visited to interview husbands in the evenings.[34] This was a "dummy operation," in the sense that the husbands were not told that the interview was in any way related to the spontaneous abortion.[35]

When Kitamura disassociated himself from the ABCC project, the spontaneous abortion project was assigned to two younger physicians, Taketa S. and Sawada H., one of whom, Taketa, had worked with Kitamura briefly in the last few months of his participation. These two continued the survey an additional two years, until 30 September 1952, but were able to collect data on only 638 early abortions in this entire period.[36] This was apparently because the community of physicians responsible for reporting the early abortions ceased to cooperate after Kitamura left. One physician interviewed by ABCC staff member Maki Hiroshi in October 1952 said that "immediately after the resignation of Dr. Kitamura . . . quite a number of people showed and had non-cooperative attitude[s] and feeling[s] toward ABCC," and he felt these feelings had affected involvement in the study. Some participating physicians also had the impression that when Kitamura left the program was discontinued.[37] Physicians responsible for about 10 percent of the total abortions reported to Kitamura dropped from the program immediately. The remaining physicians continued to be involved in the program, but their rate of reporting was greatly reduced. Taketa and Sawada were averaging only forty-five cases per month, compared to Kitamura's ninety-five.[38] In September of 1952, as a consequence

33. James V. Neel to Richard Brewer, 21 September 1950, Kitamura Program, 5/QCC, JVN.

34. Dr. Renwick, "Differences in Collection of Medical Radiation History between Kitamura, Post-Kitamura and Later Termination Programs," 14 August 1952, notebook 7, Genetics: Early Terminology, Serology, GE-41, Midwives, WJS, AA.

35. Neel to Brewer, 21 September 1950, Kitamura Program, 5/QCC, JVN.

36. Duncan McDonald, 30 September 1952, Kitamura Program, 5/QCC, JVN.

37. Maki Hiroshi to Suzuki M., "Spontaneous Abortion," 22 October 1952, Kitamura Program, 5/QCC, JVN.

38. Duncan McDonald, "Early Termination for Investigation, July 1–Dec. 31, 1951," Kitamura Program, 5/QCC, JVN.

of the previous summer's preliminary analysis of the data, the program to track early spontaneous abortion was terminated.[39]

The early spontaneous abortion data posed comparative problems that differed from every other parameter tracked in the genetics project. For most indicators, the ABCC investigators were comparing the rates of abnormal outcome in the pregnancies of two groups of parents (exposed and controls). With the early abortion data, investigators were comparing the percentage of exposed parents experiencing a particular kind of abnormality (early abortion) to the percentage of exposed parents experiencing a full-term pregnancy. For example, if one or both parents were exposed to the atomic bombs in 10 percent of the total sample of live-born offspring in a given year, but in 20 percent of those experiencing spontaneous abortion, this would suggest that survivors were disproportionately represented in the group experiencing early abortions. If atomic bomb survivors accounted for a greater percentage of the early spontaneous abortion parents than of the total parents with pregnancies carried to term, early abortion could be interpreted as a consequence of radiation exposure. The consequences of the comparison were essentially the same as those for other indicators, but the analytical implications were somewhat different.

Timing was relevant to this comparison because the percentage of parents who were survivors was constantly changing, in Hiroshima particularly. This was due to the out-migration of survivors and the high rates of in-migration of persons from other parts of Japan. The two sets of parents (full-term and early abortion) would therefore be comparable only if matched by month of conception. If the group of parents experiencing early abortion were compared to a group of parents experiencing live births even six months later, the results could suggest a spurious genetic effect, since the proportion of exposed parents in the total population was constantly falling. Conversely, if the early abortion parents were compared to a group of full-term parents who had conceived somewhat earlier, a real genetic effect could be obscured.

By the summer of 1953, many at the ABCC had reached the conclusion that the spontaneous abortion program—as important as it had often appeared to be to Neel, Schull, and others—was not working. The problems were numerous, but the most obvious was that the data collected by Kitamura over the first year of the program, and that collected by Taketa and Sawada in the second and third year, did not agree. First, the more numerous Kitamura data did not suggest any important difference in the proportion of exposed parents experiencing early abortions to those experiencing live births. For 1950 Kitamura

39. See summary of the Kitamura study and its history, James V. Neel, "Memo," 13 March 1952, 5/QCC, Kitamura Program, JVN. Also, for a memo on termination of the project, Duncan McDonald, "Memo," 30 September 1952, Kitamura Program, 5/QCC, JVN.

reported that 53.1 percent of live-born offspring had neither parent exposed, and 46.9 percent had one or both parents exposed. In the same year, 52.1 percent of couples experiencing early spontaneous abortion had neither parent exposed and 47.9 percent had one or both parent exposed—virtually the same. In contrast, the later, "post-Kitamura" spontaneous abortion data collected by Taketa and Sawada included significantly fewer exposed parents than the live-birth data. This was directly "contrary to expectation," since exposed parents should theoretically have been more prevalent among those experiencing early abortion.[40] In any case, neither data set suggested a radiation effect in the direction expected.

In both the Kitamura and post-Kitamura data, however, there was a significant excess of exposed husbands, when compared to the proportion of exposed husbands in the larger sample of live births.[41] An excess of exposed husbands suggested an effect of paternal irradiation, the kind of radiation effect Neel found most convincing. A paternal effect would be "one of our more significant findings," he noted.[42] Yet while there were proportionately more exposed husbands in the early abortion sample than in the live-birth sample, there were fewer exposed wives. The group of parents with full-term pregnancies included a higher percentage of exposed wives than did the group with early abortions. If the deficiency of exposed wives in the early abortion data set were real, Neel said, he could imagine no "logical biological explanation."[43] But he continued "to be impressed by the fact that in both sets of data we have something of an excess of exposed husbands."[44] At least the Kitamura and post-Kitamura data were consistent on this point.

These odd proportions of husbands and wives were less troubling than other discrepancies, however. For when the differences were broken down by reported symptoms and by maternal versus paternal exposure, the two sets of data diverged in ways that were almost inexplicable. The proportion of husbands reporting symptoms of radiation sickness—gingivitis and epilation— was significantly higher in the Kitamura sample, for which husbands had themselves been interviewed (not by Kitamura, but later by the genetics doctors),

40. Duncan McDonald to Research Committee, "ABCC Research Committee 1951 and 1952," 26 August 1952, drawer 3, NAS.

41. Newton Morton, "Preliminary Analysis of Early Spontaneous Abortion," 30 July 1952, 5/QCC, Kitamura Program, 5/QCC, JVN.

42. James V. Neel to Duncan McDonald, 11 September 1952. JVN.

43. James V. Neel to Duncan McDonald and Newton Morton, 12 August 1952, JVN.

44. James V. Neel to Duncan McDonald and Newton Morton, 22 August 1952, 6/Q, JVN.

than in the sample of all term parents. In the post-Kitamura data, conversely, the situation was reversed, with a much higher proportion of wives reporting having experienced gingivitis and epilation, than reported such symptoms in the sample of all term births.[45] This was the "really troublesome" aspect of the Kitamura and post-Kitamura data, Neel said. "Here we really do have a contradiction," he pointed out. The two sets of data did not agree with respect to either husbands or wives. Neel said that they might blame the problem with the Kitamura wives on Kitamura's taking casual notes in his interviews.[46] But an ABCC physician speaking directly to the husband had taken the radiation histories of the men in the Kitamura sample. In any case Neel was not entirely satisfied with the history taking explanation. "Granting the differences in the manner in which the histories were obtained . . . I find it difficult to believe that these are the entire answer."[47] In the post-Kitamura data, information on the husbands had been obtained in the usual secondhand way, from the wife. "If we say the wife's history of her husband's exposure is less accurate than that obtained directly from the husband, then all our Genetics Short Form data becomes suspect. All in all, this is just about as messy a situation as one would care to envision."[48]

These results, if taken at face value, suggested that the radiation-history taking procedures for the entire genetics project were flawed. In the post-Kitamura data, all information on exposure came from the wives, as it did in the regular genetics program. In this post-Kitamura group, there were fewer husbands with symptoms of radiation sickness than might have been expected based on the larger sample of husbands with term births. In the Kitamura data, an ABCC physician interviewed the husbands directly, and in this sample, husbands reporting symptoms were more common than in the larger sample. This suggested that reports of symptoms were more likely when the person affected (in this case, the exposed husband) was interviewed directly, then when their spouse was interviewed. Duncan McDonald had raised related questions earlier, suggesting that wives remembered the details of their husbands' exposure histories more accurately if they, too, had been exposed to the bomb. This was a particularly vexing problem if reported symptoms were to be taken as the decisive parameter for inclusion in the heavily exposed category, which Neel, Schull, Morton, and McDonald had decided they would be.

45. Duncan McDonald to Research Committee, "Early Termination Program: Exposure Categories," 26 August 1952, drawer 3, ABCC Research Committee 1951 and 1952, NAS.

46. James V. Neel to Duncan McDonald and Newton Morton, 12 August 1952, JVN.

47. James V. Neel to Duncan McDonald and Newton Morton, 22 August 1952, 6/Q, JVN.

48. James V. Neel to Duncan McDonald and Newton Morton, 12 August 1952, JVN.

As Neel, Schull, and the ABCC staff in Hiroshima debated the early spontaneous abortion program in August and September of 1952, a wide range of possible sources of bias were proposed to explain the discrepancies. Did physicians and participating hospitals in the Kitamura and post-Kitamura phases of the project differ in the relative proportion of survivors commonly treated? Did the drop in participation after Kitamura left introduce bias? How important was it that Kitamura used "on the map" (including therefore some Hiroshima suburbs) instead of "in the city" as his criteria for a survivor having been "exposed"?[49] In reviewing the many analyses and memos assessing the Kitamura data in August 1952, Neel said he felt "considerable indecision . . . as to what we should do with this whole mess of data."[50] Morton felt that it was hopeless. The data were so peculiar that the investigators could only say that "inexplicable interactions occurred between program (Kitamura and post-Kitamura) and exposure (mother and father) and between program and symptoms of mother and father." Morton could "invent no plausible biological explanation for this" and so assumed the differences were in administration or sampling.[51] McDonald's recommendation that the program be terminated closed with the conclusion that neither the Kitamura nor the post-Kitamura data could be assumed to be a random sample of all spontaneous abortions occurring in Hiroshima.[52]

By October 1952, Neel had decided that they could not salvage Kitamura's data on early spontaneous abortion. He told McDonald that the "sad demise" of this program had "one important effect" on his own thinking. "You will recall that when I was last in Japan we put together all the little hints we could assemble, and decided it just might be that we were creeping close to significant differences between the offspring of irradiated and non-irradiated parents. One of the biggest of the little hints was the Kitamura data. If we discard this— and I see no reasonable alternative at the present time—it certainly diminishes our chances of having reached a 'significant result' by the end of 1952."[53]

The Kitamura study was given half a page of discussion in the 1956 publi-

49. Eighteen similar questions, some with more details, were outlined by Newton Morton, "Biases in Early Terminations Data," 23 September 1952, Kitamura Program, 5/QCC, JVN.

50. James V. Neel to Duncan McDonald and Newton Morton, 12 August 1952, JVN.

51. Newton Morton to James V. Neel, 25 September 1952, JVN.

52. Duncan McDonald to Research Committee, "Early Termination Program: Exposure Categories," 26 August 1952, drawer 3, ABCC Research Committee 1951 and 1952, NAS.

53. James V. Neel to Duncan McDonald, 16 October 1952, JVN. "All this is valuable experience, both in how not to set up a research project and how to ferret out what actually happened and where we went wrong," McDonald noted. Duncan McDonald to James V. Neel, 3 September 1952, JVN.

cation. The differences between the first and second portions of the study were mentioned (but not detailed) and given as the reason that the early abortion study would "not be referred to again" (Neel and Schull 1956a, 18).

Neel had some reservations about the early spontaneous abortion study that were independent of this confusing breakdown of the data. He believed radiation histories taken after an undesirable pregnancy outcome were unreliable. Parents experiencing an early spontaneous abortion, he reasoned, might "remember" radiation sickness or proximity to the hypocenter as an explanation for a personal tragedy.[54] But like his scientific colleagues, he rated early spontaneous abortion as less painful for the affected parents than stillbirth or neonatal death. A reduction in fertility was not a personal tragedy in the same sense that a dead infant was. This interpretation of fertility effects was widespread in the scientific literature, articulated in many different forums and by many different individuals. As geneticist James Crow put it, a mutation resulting in "early embryonic death" may cause "little or no distress" (1957a, 71). Similarly, Earl Green noted that "in human populations, deaths of cleaving eggs or of embryos just after implantation would probably not be regarded as tragic as deaths during the infant and juvenile periods" (Green 1968, 106). And geneticist Kenneth Mather, of Birmingham University: "The loss of a potential offspring at a very early stage in foetal life might well have no effect on family size," while the "bearing of some monstrous or grossly abnormal individual might lead to the deliberate termination of child-bearing."[55] Monstrous children were a serious genetic and social problem; reduced fertility was often undetectable. Both could be manifestations of radiation-induced genetic damage, but because their social meaning differed, so too did their meaning for scientific analyses.

The question of whether a mutation caused distress was important to geneticists in general. In some ways, geneticists were concerned with measuring the social impact of increased mutation rates rather than genetic events at the level of the hereditary material. Surely the genetic events leading to spontaneous abortion were as real, in biological terms, as those leading to malformed babies. But the important measures focused on social cost, on the implications for human suffering, and on the possible damage to the gene pool. These priorities led Neel and Schull, and their advisers and collaborators, to dismiss another possible sign of genetic change for which there were justifications from both mouse and *Drosophila* studies. This was minor malformation.

54. James V. Neel, interview by author, 20 November 1988, Ann Arbor, Michigan.

55. Kenneth Mather, "The Long-Term Genetical Hazard of Atomic Energy," in Haddow (1952, 57–66).

MINOR MALFORMATION

Throughout the study, and particularly in 1952 and 1953 when they prepared the preliminary analysis, Neel, Schull and the ABCC staff pondered the question of what should be classified as a major malformation and what as a minor malformation.[56] "Major" and "minor" seem to describe the degree of medical severity of a condition, but the terms were also (perhaps primarily) categories for deciding what counted in the analysis. Abnormalities designated as major were included in the tabulations; those designated as minor were not.[57]

The four articulated criteria for determining a trait's classification were suspected cause (nongenetic or genetic), difficulty of diagnosis in a newborn (not easily detected, easily detected), frequency (rare or common), and health implications (trivial or serious). The major/minor dichotomy was most consistent in relation to abnormalities that were suspected to be nongenetic, easily detected, common, and without serious medical implications (certain birthmarks, for example). These would generally be classified as minor malformations. Congenital anomalies recognized as major from the beginning of the study included polydactyly (extra fingers or toes), syndactyly (webbed or fused fingers or toes), harelip, cleft palate, undescended or absent testes, anencephaly (lack of brain), hydrocephalus (water on the brain), and icthyosis congenita (a rough, scaly skin condition).[58] Those classified as minor included the presence of teeth (or a single tooth) in a newborn; various birthmarks, including the "stork bites" commonly found on the back of the neck; malformed ears; umbilical hernias; partial albinism; and moles.[59]

Sometimes difficulty of diagnosis was the primary criterion for designating a particular trait as minor. If a pediatrician could easily overlook a trait in

56. See particularly James V. Neel to Duncan McDonald, 23 June 1952, JVN; also, John W. Wood, "Diagnostic Criteria for Congential Malformations of the Heart," 29 July 1952, notebook 9, Genetics Long Form, WJS, AA.

57. Neel told the ABCC pediatricians to "record everything" in their home visits. But he began to doubt the value of coding "minor" abnormalities as early as December 1948. By the summer of 1949 Neel decided minor malformations were a diagnostic problem. Standards of reporting varied among the three cities. Hiroshima and Kure physicians tended to record every malformation, even "very minor" ones. This attempt Neel estimated as about 80 percent effective. "There was, if anything, a tendency to 'over-diagnose' on the part of certain interns." On the other hand, in Nagasaki, many minor abnormalities, such as birthmarks, were not reported. Accuracy, in the sense of "recording everything," was closer to 10 or 20 percent in Nagasaki. James V. Neel, "Report to the Committee on Atomic Casualties, 8 May 1949 to 13 July 1949," JVN. By June 1952, Neel decided that the data on minor malformation should not be used. James V. Neel to Duncan McDonald, 23 June 1952, 6/Q, JVN.

58. "Summary of 1948 Congenital Abnormalities, Major," 5/BB, JVN.

59. "Summary of 1948 Congenital Abnormalities, Minor," 5/BB, JVN.

the examination of a newborn, it was often classified as minor. This was done in the belief that rates of diagnosis for difficult-to-detect conditions would be inconsistent from doctor to doctor. At one point Neel suggested that all heart anomalies be coded as minor because of this diagnostic problem, but this was resisted by John Wood, a physician then working in Hiroshima, and in the end Wood's classification scheme was accepted.[60] Wood said heart murmurs in newborns were "of questionable significance" but that if the murmur persisted for several weeks, it was probably a sign of a congenital malformation of the heart. All infants with heart murmurs detected in the at-home examination (during the first four weeks of life), Wood said, should be brought in to the ABCC clinic "for verification by an Allied pediatrician"; those exhibiting both a murmur and cyanosis could be classified as having a major abnormality.[61] This meant that a good number of major malformations coded on early reports, prior to Wood's clarification in 1952, had to be recoded as minor because insufficient information was available.[62]

Stanley Wright, a physician working in the Nagasaki ABCC, attempted to recheck all cases of congenital heart disease in the Nagasaki sample. Many of those never seen at the clinic, that is, those examined only by Japanese pediatricians in the home visits, had to be classified as minor, he reported, because "the doctors did not grade the murmurs and consequently it is difficult to evaluate their description." He suspected that they were actually all major, because "when they were able to hear a murmur it had to be quite loud." But the majority of those seen in the nine-months examination did meet Wood's criteria for majors. Wright said Wood's classification scheme was "an improvement over

60. Neel made the decision that they should "code all hearts as minor." One American pediatrician then working in Japan, John Wood, argued against such a classification, and Neel finally gave in. "I think we can retain those hearts as major which meet his [Wood's] specifications." John Wood, interview by author, Ithaca, New York, March 1988; James V. Neel to Duncan McDonald, 23 June 1952, JVN.

61. John W. Wood, "Diagnostic Criteria for Congenital Malformations of the Heart," 29 July 1952, notebook 9, Genetics Long Form, WJS, AA.

62. Stanley Wright, a physician working in Nagasaki ABCC, attempted to recheck all cases of congenital heart disease in the Nagasaki sample. Many of those never seen at clinic, that is, those examined only by Japanese pediatricians in the home visits, had to be classified as "minor," he reported, because "the doctors did not grade the murmurs and consequently it is difficult to evaluate their description." He suspected that they were actually all "majors," because "when they were able to hear a murmur it had to be quite loud." The majority of those seen at clinic (in the nine-month examination) did meet Wood's criteria for "majors." Wright said Wood's classification scheme was "an improvement over the previous one," in which all heart anomalies were majors, but it was "not clinically a satisfactory grouping" since "an infant with a serious form of heart disease, potential or real, might change from a 'minor' to a 'major' and back again, even in the same day." Stanley Wright to Duncan McDonald, 30 April 1952. unlabeled notebook, WJS, AA.

the previous one," in which all heart anomalies were majors, but it was "not clinically a satisfactory grouping" since "an infant with a serious form of heart disease, potential or real, might change from a 'minor' to a 'major' and back again, even in the same day."[63]

Frequency, medical severity, and difficulty of diagnosis all figured into the classification of an abnormality as major or minor, though not necessarily in a consistent manner. A relatively serious condition that was easily diagnosed, but believed to be nongenetic in origin, might be classified as minor in order to remove it from the analysis. Traits that were "obviously of little importance" genetically were removed from the category of "major malformation" even if they constituted a health problem. Until the summer of 1952, for example, the ABCC staff was coding pilonidal sinuses—ingrown hairs in the buttocks—as major malformations. McDonald objected that while this made sense "from the point of view of the health of the child," it did not "from an embryological and genetic aspect." Pilonidal sinuses, he said, were nongenetic in origin and should therefore be coded as minor.[64] The questions about pilonidal sinuses led Neel to question two other major malformations that were "less major than the other majors," namely, dysplasia of the acetabulum (malformed hip socket) and inguinal hernia. Perhaps these two should be treated in a "somewhat differ-ent manner from the other major malformations," partly because they were so frequent.[65] In the end inguinal hernia was treated differently depending on sex; it is common in boys but uncommon in girls, so was classified as minor in boys but major in girls.[66] (Indeed, for reasons that remain unclear to me, Neel generally interpreted frequently appearing traits as less important.)[67] Some se-

63. Stanley Wright to Duncan McDonald, 30 April 1952, unlabeled notebook, WJS, AA.

64. Duncan McDonald to John W. Wood and James V. Neel, "Coding of Pilonidal Sinuses," 13 August 1952, notebook 9, Genetics Long Form, WJS, AA. Also, Duncan McDonald to James V. Neel, 21 May 1952, 6/T, JVN: "Pilonidal sinuses still worry us; these have been recorded as majors but genetically are so obviously of little importance that it scarcely seems reasonable to include them with the other majors."

65. James V. Neel and William J. Schull to Duncan McDonald and Newton Morton, 30 August 1952, 6/Q, JVN.

66. Inguinal hernias in males appeared often in the ABCC sample, were relatively easy to diagnose, and were relatively serious. Inguinal hernias are small protuberances of intestinal mate-rials through the inguinal ring, the opening in the musculature of the pelvis through which the sex organs drop during fetal development. See James V. Neel to Duncan McDonald, 23 June 1952, 6/T, JVN.

67. Neel also questioned two major malformations that were "less major than the other ma-jors," namely, dysplasia of the acetabulum (malformed hip socket) and inguinal hernia (hernia in the muscles supporting the viscera). Perhaps these two should be treated in a "somewhat different manner from the other major malformations," he suggested, partly because they were so frequent.

rious malformations were classified as minor because they were difficult to diagnose in the newborn. Other traits that were difficult to diagnose but extremely rare were classified as major and therefore included in the analysis.

Neel told the ABCC pediatricians to "record everything" in their home visits. But he began to doubt the value of coding minor abnormalities as early as December 1948. By the summer of 1949 Neel decided minor malformations were a diagnostic problem. Standards of reporting varied among the three cities. Hiroshima and Kure physicians tended to record every malformation, even "very minor" ones. This attempt Neel estimated as about 80 percent effective. "There was, if anything, a tendency to 'over-diagnose' on the part of certain interns." On the other hand, in Nagasaki, many minor abnormalities, such as birthmarks, were not reported. Accuracy, in the sense of "recording everything," was assessed at closer to 10 or 20 percent in Nagasaki.[68] By June 1952, Neel decided that the data on minor malformation should not be used.[69]

Many of the abnormalities classified as minor appear with some frequency in human populations. Including them in the analysis would have resulted in a larger data base when comparing exposed and control groups. In 1948, for example, there were 59 major malformations coded in Hiroshima, Nagasaki, and Kure, out of 6,520 births; for the same year, in only Hiroshima and Kure, there were 253 minor malformations out of 5,808 births.[70] Among newborns examined in Hiroshima in 1949, major abnormalities afflicted 1.18 percent of control offspring, 1.05 percent of the offspring of exposed fathers, 1.78 percent of the offspring of exposed mothers, and 1.29 percent of the offspring of two exposed parents. Minor abnormalities were diagnosed in 14.02 percent of those with neither parent exposed, 15.14 percent with father only, 12.96 percent with mother only, and 13.57 percent with both parents exposed. These numbers cannot be taken as evidence either for or against exposure effects for several reasons, including the fact that those counted as exposed in this early calculation included large numbers of survivors who were distant from the hypocenter, and who were later classified as controls because of their relatively low radiation exposure. But the numbers do indicate the frequency of minor malformation.[71] By 1952, the majority of children examined by the ABCC

James V. Neel and William J. Schull to Duncan McDonald and Newton Morton, 30 August 1952, 6/Q, JVN.

68. James V. Neel, "Report to the Committee on Atomic Casualties, 8 May 1949 to 13 July 1949," JVN.

69. James V. Neel to Duncan McDonald, 23 June 1952, 6/Q, JVN.

70. "Summary of 1948 Congenital Abnormalities," 5/BB, JVN.

71. See George A. Hardie, "Report of a Visit to ABCC in Japan, June–July 1950 [Copy No. 3]," ABCC, NAS, 79.

were "found to have some minor aberration or other at the age of 9 months." To include them in the analysis would therefore "involve in the neighborhood of 50 percent of all children seen at age 9 months."[72]

Neel and Schull have explained that they rejected minor malformations because such malformations were difficult to diagnose and because they did not know the degree to which such minor malformations were under genetic control.[73] I would suggest that the rejection was also a consequence of their understanding of the concept of mutation. The objections they raise to minor malformations also apply to many of the major malformations included in the analysis. While some of the malformations classified as minor were difficult to diagnose, others were obvious and easily diagnosed. The partial albinism in which an otherwise normal Japanese baby was born with white hair, for example, was unmistakable. And pediatricians could easily miss some of the abnormalities listed as major, such as inguinal hernias in females. Degree of genetic control, similarly, was to some degree problematic for virtually all the indicators, not just those classified as minor. The genetic sensitivity of most of the conditions chosen for analysis was unknown. And some traits known to have a large environmental component, such as birth weight and growth, counted in the ABCC analysis.

The minor/major dichotomy does not break down consistently into traits of major medical importance and traits of minor or no medical importance. But the dichotomy does seem relatively consistent when viewed in terms of social consequences. While the reasons for assigning a particular trait to either category were complex, minor malformations were generally those with minimal social implications for the affected infant or for the human gene pool. The decision to exclude minor malformation was consistent with the related decisions to pay less attention to reduced fertility and to early spontaneous abortion—because they had a less direct impact on both "human suffering" and human evolution. The decision to exclude minor malformation from the analysis provides insight into the meaning that mutation had in practice.

CONCLUSION

Geneticists working with flies or mice considered even slight variations to be products of induced mutation. Minor differences counted: white-eyed flies

---

72. James V. Neel and William J. Schull to Duncan McDonald and Newton Morton, 30 August 1952, 6/Q, JVN.

73. James V. Neel and William J. Schull, interview by author, March 1988, Hiroshima, Japan.

could be white, ecru, buff, ivory; red could be coral, blood, cherry; and any of these variations would be classified as a mutation if appearing as a new trait. Similarly a new trait in a mouse would be tabulated as a mutation even if it were insignificant—a slight variation in ear shape, nostril configuration, or coat consistency, for example.[74] But investigators working with human populations did not count the human equivalent of a variation in eye shade or wing shape as a mutation. Strangely shaped ears or changes in shades of hair or eye color are not socially or medically disadvantageous. They are therefore not mutations.

This is in many ways a legitimate distinction. Standards for human populations are different, and a slight shift in hair or eye color as a consequence of mutation, for example, does not have any particular importance either to the mutant individual or to human society in general. But these differing standards may have consequences for cross-species comparisons of mutation rates and radiation sensitivity.

The ABCC study focused on a different kind of event from that measured in laboratory studies with experimental organisms. The ABCC geneticists were neither counting the same kinds of traits nor using the same meaning for mutation (Neel and Schull 1956a, 205–17). If a human mutation was a deleterious change with implications for the human future, then minor malformation, sterility or reduced fertility, or even early spontaneous abortion were not signs of mutation. They did not add anything to the load of the human gene pool. They were therefore not relevant to the problems posed by human exposure to ionizing radiation.

While Neel does not believe that political concerns played any role in the selection of what signs counted as mutations at Hiroshima and Nagasaki, he acknowledges that social impact did play a role. A trait that would not "upset the new mothers of Hiroshima," he has explained, should not count as a mutation.[75] Such a formulation interprets mutation as a distressful change. In an interview in 1988, Neel proposed poor academic performance in elementary school as a possible sign of mutation among the offspring of atomic bomb survivors.[76] In terms of known degree of genetic control, the mediating factors between mutation and elementary school performance are at least as complex as those between mutation and minor malformation of any kind in a newborn. But the differences between the two are suggestive. Birthmarks are not socially

---

74. For a review of mouse literature, see A. G. Searle, "Mutation Induction in Mice" (1974). Also, W. L. Russell, "Comparison of X-Ray-Induced Mutation Rates in Drosophila and Mice" (1956).

75. James V. Neel, interview by author, 20 November 1988, Ann Arbor, Michigan.

76. James V. Neel, interview by author, 22 November 1988, Ann Arbor, Michigan.

or politically dangerous; poor school performance is perceived as a social problem with wide-ranging implications.[77] The relative tragedy of a trait—its contribution to human misery—mattered to Neel and Schull and their fellow geneticists. A mutation in this data set was a dangerous, threatening, or socially disturbing trait with implications for future human survival. This was the chosen scientific understanding of mutation in human populations partly because this was the concept prevalent in the larger political debate.

The working definition of a scientific parameter here reflected the context within which these scientists expected their results to be received. A "change in the hereditary material" is sufficiently ambiguous to permit several interpretations. As my study suggests, the exploration of such scientific concepts and their use in the construction of a particular data set can be a productive way to understand how cultural and social beliefs are incorporated into scientific practice.

While various explanations were devised for particular decisions about major and minor malformations, participants shared the fundamental underlying assumption that social impact mattered. In cases lacking any compelling reason to decide otherwise, traits that were socially dangerous were classified as major and those without social implications as minor—essentially, as not mutations.

In my next chapter, I explore the first comprehensive working analysis of the genetics data, an analysis which led to the termination of the morphological genetics program in early 1954.

77. The problems of the American school system have been contentious for most of this century. See Lawrence Cremin, *The Transformation of the School* (1964), and National Commission on Excellence in Education (1984).

CHAPTER TEN

Draft Analysis, 1952–1953

By 1952, when the first serious analysis of the ABCC genetics data began, the raw numbers were uninformative. A simple comparison of the indicators in the various exposed and control groups did not suggest any coherent trends. For some indicators, the results suggested a slight genetic effect in the direction expected. Sex ratio appeared to be moving toward significance in the early results and continued to be suggestive if not conclusive.[1] Stillbirth rates for Nagasaki survivors seemed to be higher. For other indica-

1. In the preliminary report of the ABCC genetics results, Neel, Schull, McDonald, Morton, and others reported a significant sex-ratio effect (Neel, Schull, McDonald, et al. 1953b; see also Schull, Neel, and Hashizume 1966). In the 1956 monograph, they noted that the effects seemed to have diminished but that the trends were still consistent with sex-ratio effects of the kind to be expected (Neel and Schull 1956a).

tors, the numbers were uninformative. And some data suggested radiation could have beneficial effects: birth weight for children of survivor parents was slightly higher.

Taken simply in the aggregate, the ABCC genetics data revealed very little. It suggested at least that radiation had not produced a dramatic (for example, more than 100 percent) rise in the rates of abnormal births to exposed parents, but even reaching this conclusion would require some further processing of the data. Neel and Schull needed to control for confounding variables, to minimize the impact of uncertainties over dosimetry, and to break down the data in small clusters that might provide more powerful insight into the differences in pregnancy termination among exposed and control groups. They faced the difficult task of deriving a productive comparison from problematic data that emphatically did not speak for themselves.

Their process of making statistical sense of their numbers and intellectual sense of their assumptions and their ideas is the subject of this chapter. The analysis I examine here is not the final analysis presented in the "blue book" of 1956, *The Effect of Exposure to the Atomic Bombs on Pregnancy Termination in Hiroshima and Nagasaki,* the major publication on the genetics study. I explore that presentation and analysis as public strategy in chapter 11. Here I consider instead the working analysis carried out in 1952 and 1953 to determine the usefulness of continuing the genetics project after 1954. This analysis suggested radiation effects would remain undetectable even if the ABCC were to continue to collect data for several more years. It led to the recommendation by a National Academy of Sciences genetics conference in the summer of 1953 to terminate the morphological genetics program by June 1954. In fact the program was terminated early, at the end of January 1954. The collection of data on the sex ratio, which was the only parameter suggesting positive results at the statistically significant level, was continued. But other genetics data gathering ceased. The analysis in 1952 and 1953 was therefore more important, in terms of the fate of the project, than the later published analysis, which differed in several ways from the earlier version. The preliminary analysis or "pilot program" of this period effectively ended the laborious examination of every child born in Hiroshima and Nagasaki.

Neel and Schull told a group called together to review the genetics program in 1953 that the analysis suggested no significant differences between heavily exposed and nonexposed or lightly exposed control groups for malformation, spontaneous abortion, neonatal mortality, or growth at nine months. There were, they reported, significant differences in type of termination (stillbirth, term), sex ratio, and birth weight. These differences were not all in the expected direction: birth weight was higher in the offspring of survivor parents, though radiation exposure had been expected to result in a lower birth weight.

There seemed to be a slightly greater frequency of stillbirths in heavily exposed mothers but there was no similar effect from paternal exposure. The
clearest difference was in sex ratio: the proportion of males born to heavily
exposed mothers was 50.16 percent, compared to 51.92 percent for control
mothers. This variation was in the direction expected and was the only finding
that seemed to clearly indicate a radiation effect.[2]

Much of the analysis that formed the basis of this report was worked out
in the 1952 and 1953 correspondence between James Neel, William Schull,
Newton Morton, and Duncan McDonald. Neel and Schull (both in Ann Arbor,
Michigan) debated by trans-Pacific mail with population geneticist Morton and
eccentric Canadian biologist McDonald (both in Hiroshima). McDonald was
in some ways the outsider, trained as a biologist rather than a geneticist and
not destined for a career in academic research science.[3] Morton, a bright young
statistical adventurer hoping to develop a dissertation based on his work in
Japan,[4] would soon return to Madison to work with James Crow. Schull and
Neel were clearly the most directly engaged, and they often, though not always,
prevailed over their correspondents in Hiroshima. Over the course of the debate, these four tackled questions about how to set up the exposure classes and
about what information should be given priority in the radiation histories. They
also tried to compensate for differences in age, number of pregnancies, cousin-
marriage, and other confounding variables among the groups they were comparing.

Morton sometimes bemoaned the difficulties of debating scientific points
by mail. "It is too bad that you aren't here or we aren't there—one conversation
would very likely eliminate this barrage of letters crossing in the mail."[5] Yet
both Neel and McDonald pointed out that the letters gave them a written record
of how decisions were reached.[6] The letters even provided an intellectual his-

2. "Statement of the Conference on Genetics to the Committee on Atomic Casualties," 10–11
July 1953, 6/R, JVN.

3. McDonald sometimes questioned the statisticians: "Dr. Brandt is much against known biases, even in the right direction—if you know it, get rid of it." Duncan McDonald to William J.
Schull, 14 December 1951, unlabeled notebooks, WJS, AA.

4. Morton was a 1951 graduate of the University of Hawaii, and a 1952 graduate with a master
of science degree from the University of Wisconsin, where he had worked with James Crow. His
tenure at the Atomic Bomb Casualty Commission lasted only two years, 1952 and 1953, after
which he returned to the University of Wisconsin, completing his Ph.D. in genetics in 1955. That
same year Morton published an analysis of the inheritance of human birth weight that was based
partly on his work with the Atomic Bomb Casualty Commission (Morton 1955).

5. Newton Morton to James V. Neel, 21 August 1952, JVN.

6. Schull became the group's memoirist. His account of the ABCC, *Song Among the Ruins,*
was published in 1990. Schull was asked by Grant Taylor in early 1952 to prepare a monograph

tory of their disagreements. "We shall have a permanent record of how the final analysis was arrived at. Personal talks would cut down on time, but we might forget why we decided on any given step."[7] Neel regretted his failure to keep a comprehensive diary of his work with the ABCC, but he told McDonald (correctly) that their correspondence would be useful to future historians.[8]

As Neel, Schull, and their collaborators were aware, there were many reasons why a real effect might be missed under the circumstances of the ABCC study, including uncertainties about dosimetry and the vagaries of taking radiation histories. At the same time, there were many reasons why a spurious genetic effect might be indicated, as a possible artifact of the data-gathering process or the nature of the information available.

It was possible to compare the data in ways that revealed significant genetic effects, and it was equally possible to compare the data in ways that revealed no genetic effects or even beneficial effects. Many of the choices and decisions that would lead to one or the other conclusion depended on judgments about what problems were more important, for virtually all the data collected in this ABCC study could be termed problematic. Ideas about human nature, about what people are likely to reveal and what they are likely to conceal, helped determine the degree of faith placed in some of the data. Perceptions of the relative risk of making a given pronouncement—some findings were clearly much more likely to provoke public or scientific response—affected the degree of evidence that the investigators felt would be necessary to support the statement. Deciding how seriously a particular confounding variable should be taken was often a matter of "common sense." How much weight should be given to extrapolation from the mouse data? What indicators, under what circumstances, were more convincing than others? And what differences in the control and exposed populations mattered *most* (for there were many differences)? Members of the ABCC team did not always agree among themselves about such questions. In many of their disputes, it was possible to argue the case either way, as their debates indicate, based on the level of scientific

---

on the history of the ABCC. He responded to that proposal as follows: "There are too many people who would be unhappy at having the errors they committed in the management of this program reviewed in such a manuscript. I can't visualize Dr. Shields Warren or Dr. Detlev Bronk, for example, as being particularly happy with the publication of a monograph which could reflect upon their role in connection with ABCC's history. I'm afraid this would become a political 'hot potato' which might jeopardize all of the good work that has gone on in Japan." He closed by saying that he felt that this story should eventually be told, but "I'm equally convinced that now is not the best time to do it." William J. Schull to Grant Taylor, 21 January 1952, folder 2, WJS, HAM.

7. Duncan McDonald to James V. Neel, 22 August 1952, JVN.

8. James V. Neel to Duncan McDonald, 16 October 1952, JVN.

knowledge at the time. The choices made depended on perceptions about what problems mattered more. While those perceptions were often grounded in scientific norms, they also turned on what might be called aesthetic judgments about what data were more satisfying or relevant.

In this chapter, I examine two debates in the preliminary data analysis in 1952 and 1953. I consider, first, how the participants chose to deal with dosimetry and, second, how they weighted differences in their exposed and control populations. Their resolution of these analytical problems suggests the priorities and concerns that shaped the preliminary results in 1953.

DOSIMETRY AND THE CLASSIFICATION OF PARENTAL EXPOSURE

There was one fundamental uncertainty that affected all the genetics data: How much radiation had each survivor absorbed? All the other complications and confounding variables in the analysis of this data were minor by comparison. The geneticists could not themselves solve the problem of dosimetry—the physicists were still working on the problem, and the details were still classified in 1952. Neel, Schull, and their collaborators needed to know how much radiation had reached the gonadal cells of those exposed. But an answer would have required detailed knowledge of both the amount and types of radiation released by the two bombs (they differed, due to different construction and different cores, and their radiation mixes were classified) and the exact location, body position and shielding of each survivor. Ideally they would have calculated precisely each parent's gonadal level of radiation exposure, and thereby classified each pregnancy based on the total exposure of the two parents. But such a calculation was impossible, and indeed, uncertainties about dosimetry persist into the present, despite repeated and increasingly sophisticated efforts to assess the exposure level of each survivor.[9]

Uncertainties about dosimetry threatened to limit the usefulness of the study as a whole, since one important purpose of the study was to suggest acceptable levels of radiation exposures for those occupationally or medically exposed to low levels of radiation. Low-dose effects were more important, in this context, than high-dose effects.[10]

9. On the new dosimetry estimates for the atomic bomb survivors, see Roesch (1987).

10. As Karl Morgan noted, commenting on the ABCC results, "Whether exposure of occupational workers or of a population [exposed to a reactor accident] is considered—perhaps even in the case of atomic war—those receiving low exposure may hopefully be the vast majority of cases. Thus even though the doses may be small, the greatly increased number of persons potentially in this range makes the knowledge concerning effects at low dose levels of great importance." Karl

In terms of radiation genetics, the exposure levels of the most heavily exposed survivors who participated in the genetics study were relatively low. Most of those exposed to 600 roentgens of gamma rays or more did not survive beyond the first few weeks after the bombings. In addition, in order to be included in the genetics study, survivors had to reproduce. Those severely debilitated by the bombings—as a consequence of radiation exposure or other injury—were presumably less likely to later become parents. The genetics study therefore necessarily focused on a small subset of survivors, generally consisting of those least affected by the bombings and most capable of carrying on with their lives.

Some criticized the ABCC studies of the survivors on this basis. British scientist Alice Stewart argued, for example, that the survivors, having endured a selection bottleneck, were an unusually robust population and therefore could not provide insight into the effects of radiation exposure on populations in general. In her 1982 assessment of the "selective survival of exceptionally fit individuals," Stewart invoked a "silent forces" hypothesis. She said that the relatively low rates of general mortality in bomb survivors was a consequence of their "exceptionally high rates [of mortality] during the immediate aftermath of the explosions." Anyone already in a poor or even mildly impaired state of health, she argued, was more likely to die during the general devastation and chaos that followed the bombings. ABCC epidemiologist Gilbert Beebe acknowledged this possibility in a 1978 summary report of the ABCC results (Beebe, Kato, and Land 1978b), and it is possible that this differential survival did affect the legitimacy of extrapolating from the Hiroshima and Nagasaki data to other populations. A more "robust" population might also be considered more genetically healthy and therefore not a suitable test population for the genetic effects of radiation.

But this was not a parameter over which the ABCC geneticists had any control. The geneticists studying the survivors could not possibly assess the total mutational impact of exposure to the bombs; this was a simple and unavoidable practical constraint, present in virtually all studies of radiation impact, including those with model organisms. They could only look for patterns in the offspring of those survivors who were both of suitable age and suitable health to reproduce within the five years of the morphological study, 1948 through 1953.

This group was studied with the express purpose of gaining insight into acceptable levels of radiation exposure for those occupationally or medically exposed to chronic, low levels of radiation. Dosimetry was therefore quite im-

---

Z. Morgan to George Darling, 1 July 1963, drawer 4, Oak Ridge Correspondence, General, 1957–1967, NAS.

portant. If radiation protection standards were not a central issue—if the questions to be answered were simply yes-no questions about genetic effects—the survivors might have simply been categorized in exposure groups (heavy, light) without an attached numerical value (presumed level of radiation exposure). This would have allowed the ABCC team to avoid statements about dosimetry until more information was available. But such results would have been of far less value both politically and scientifically. Without dosimetry estimates the results could not be used to set exposure limits, nor could they be compared to results with mice or *Drosophila*. The various purposes of the research demanded that some quantitative assessment of radiation exposure be undertaken even though dosimetry was unquestionably the most complicated and difficult unknown factor in the study.

Many of the decisions made about shielding and dosimetry in the early years of the ABCC were later found to be sources of error. The ABCC staff assumed, for example, that the proper way to calculate shielding was to draw a straight line from the survivor to the epicenter, that is, the in-air point of explosion of the bomb. Any materials in this line were then assessed as having the shielding properties of water and the dose was calculated based on an equivalent level of protection by water.[11] Later investigators concluded that the survivors, as a consequence of the scattering of radiation, would have been exposed from all directions, rather than only along a line from the burst point. The protective value of various shielding materials also varied greatly: water was not a suitable standard. In addition, the early estimates of the radiation released by the two bombs were later found to be too high. This meant that any effects detected in the populations at Hiroshima and Nagasaki were the product of much lower radiation exposures than originally believed. (Recently, however, a biophysicist at Lawrence Livermore National Laboratory has suggested that the 1986 recalculation of radiation dosimetry at Hiroshima seriously *underestimated* the amount of radiation released by the bomb; thus the issue is far from being resolved even in the 1990s [Marshall 1992, 394].) The dosage estimates used by the ABCC and in the original report of the genetics study were "completely inadequate and misleading," not through oversight on the part of the ABCC investigators, but as a consequence of the information then available about the bombs' radiation and the nature of shielding and radiation diffusion (Auxier 1977, 20).

There were also persistent uncertainties about residual radiation, that is, radiation remaining in the area around the hypocenter after the bombings. Jap-

11. This equation is detailed in Oughterson and Warren (1956). For a discussion of its impact on the efforts to determine dosimetry, see John Auxier, *Ichiban: Radiation Dosimetry for the Survivors of the Bombings of Hiroshima and Nagasaki* (1977, 4, 20–21).

anese surveys indicated that residual radiation might have been a problem for the "early entrants," rescue workers who came into Hiroshima or Nagasaki shortly after the bombings. In 1950 another Japanese survey in Hiroshima and Nagasaki found traces of residual radioactivity in soil samples sufficiently high to fog a sheet of film when the samples were placed in contact with the film for prolonged periods. While this was undoubtedly a very low level it was difficult to assess the biological impact it might have in the long term. ABCC director Carl Tessmer, at least, felt the results were significant enough to call for a reassessment of the decision that year to abandon Kure as a control city.[12]

I stress these problems with dosimetry because I want to suggest that a highly problematic dimension of a data set can cease to be an issue in scientific debate precisely because it is such a serious problem. Dosimetry was an unknown that was out of the control of Neel, Schull, Morton, and McDonald. They were forced to do what they could with shielding, distance, and symptoms, and then assume that dosimetry was more or less adequate, because so little was known as to preclude more serious debate. Neel commented in 1953 that, in retrospect, he could not "help but wonder whether the study would have been undertaken at all if as much were known then as is now known about the dosage factor."[13] He and his colleagues worked from the assumption that any errors in dosimetry would not negate the value of the results. They knew all too well that dosimetry was a serious problem, but they could not solve it and it therefore ceased to be an issue. In the end, they made some choices that put more weight on the dosimetry classifications than on other kinds of data that were at least no more problematic, and sometimes less problematic, than the dosage estimates. The problem of dosimetry in the genetics study in 1952 and 1953 was not resolved by more conclusive information, or by greater precision in determining the position and exposure levels of individual survivors. It was rather resolved by the four participants—Neel, Schull, Morton, and McDonald—negotiating an acceptable level of uncertainty that prevented dosimetry from interfering with other analytical choices and decisions.

In 1952, assigning individual doses to the survivors was impossible. Many of the radiation exposure estimates in the early ABCC studies were drawn from a report describing the radiation released by a "typical" atomic bomb, and neither the Hiroshima nor the Nagasaki bomb was exactly typical.[14] The ABCC team also had access to a report prepared for the Committee on Atomic

12. Carl Tessmer to Philip Owen and Herman Wigodsky, 29 March 1950, drawer 3, NAS.

13. James V. Neel to Duncan McDonald and Newton Morton, 15 June 1953, notebook 5, Genetics, Correspondence II, WJS, AA.

14. The "typical" bomb was described in Samuel Glasstone, *Effects of Atomic Weapons* (1950). Glasstone presented the effects of a "nominal" atomic bomb, defined as one whose total

Casualties by Robert R. Wilson, which estimated levels of various types of radiation at Hiroshima and Nagasaki.[15] While individual doses for each parent would have allowed a more productive regression analysis, Neel, Schull, and their collaborators had to settle for assigning survivors to groups, that is, lumping together those who seemed to have received similar levels of radiation exposure. They could choose to recognize any number of groups. The simplest option would be to designate parents as exposed or not exposed. The investigators could then compare the pregnancy outcomes in the four resulting classifications (both parents exposed, neither parent exposed, mother only exposed, father only exposed).

This seemed too crude, however, and Neel and Schull chose instead to classify the parents according to five levels of exposure. Parents in category 1, the un-exposed group, had been outside the cities at the time of the bombings. Parents in category 5, the most heavily exposed, had been near a hypocenter and reported experiencing symptoms of radiation sickness, including hair loss, nausea, subcutaneous bleeding, or gingivitis. The three classes in the middle (2, 3, and 4) consisted of those exposed to the bomb at varying distances from the hypocenter, with varying levels of shielding, who did not report symptoms of radiation sickness. A pregnancy would therefore be classified on the basis of combined maternal and paternal exposure in one of twenty-five possible combinations.

Classifying the parents into any group was quite difficult, however. When an informant's story contained internal inconsistencies the scientists disagreed about what information should be given precedence. For example, when a survivor said she had been distant from the hypocenter, but reported experiencing severe radiation sickness, she could conceivably be placed in one of two different categories. She could go into the category consistent with her reported location (2, 3, or 4) or she could go into the most heavily exposed category (5) based on her reported symptoms. Relatively few of those interviewed in the genetics program fell into this ambiguous category. Most of those who experienced symptoms also reported their location as within twenty-five hundred meters of the hypocenter at the time of the bombing. But Neel estimated that "about 6 percent of those who present significant symptoms were at a distance of more than 2,500 meters from the hypocenter." Roughly 80 percent of this group was in the region between twenty-five hundred and twenty-six hundred

---

energy release was the same as twenty thousand tons of TNT. This proved to be a fairly good estimate for the Nagasaki bomb and a poor one for the Hiroshima bomb (Auxier 1977, 6).

15. This report was prepared in 1950 and presented to the Committee on Atomic Casualties of the NAS in 1951. It was declassified in 1955 and published (in amended form) in 1956 (Wilson 1956).

meters, still quite close to the hypocenter.[16] The number of individuals involved in this question was, then, rather small. Yet the scientists' conclusions about what information to trust in such cases suggests how ideas about human nature shaped their negotiation and construction of the data analysis.

Morton felt that reported distance was more reliable than reported symptoms, and survivors should be classified strictly by their self-reported location at the moment of the explosion. Symptoms should be "interpreted more conservatively" (that is, mistrusted more) than reports of distance or shielding.[17] Neel took the opposite position, arguing that informants' reports of symptoms could be better trusted than their reports of location. McDonald sided with Morton,[18] Schull sided with Neel, and the debate raged on for some weeks, with much anecdotal evidence proffered by both sides. Neel said he felt "reasonably sure" he would remember being sick two weeks after the bombing better than he would remember his exact location at the moment of the bombing, "when I was frightened almost to death."[19] McDonald countered that he had it personally from "an RCAF chap who was here some months ago" that at the "moment of supreme disaster" one remembers "every detail." "'Engraved on my mind still' was his expression" in describing his experiences during the 1923 earthquake in Japan, McDonald said. And he had heard the same from "London bomb victims."[20]

Schull appealed to quantification. He had earlier prepared a study of women who had registered two pregnancies with the ABCC. Attempting to assess the validity of the radiation histories, he compared these women's stories of exposure in their first interview with their stories in their second interview. He found 188 sets of histories that did not entirely agree. Of these, 110 informants had changed their distance from the hypocenter. Forty-six had changed from being present in the city to absent, or vice versa, and thirty-two had changed their age. But there was "*not a single change in symptoms*" (emphasis in original). Schull, recalling this earlier study, told McDonald that this did not mean that survivors never changed their reports of symptoms—his sample was small—but he did interpret it as meaning that reports of symptoms were more

16. James V. Neel and William J. Schull to Duncan McDonald and Newton Morton, 30 August 1952, 6/Q, JVN.

17. Newton Morton to James V. Neel, 1 August 1952, JVN.

18, Morton said "estimates for symptoms can only be interpreted as applying to a group of people who under certain ascertainment procedures give positive answers, a somewhat ill-defined population; presumably distance is a more reliable criterion." Newton Morton to James V. Neel, 14 August 1952, JVN.

19. James V. Neel to Newton Morton and Duncan McDonald, 12 August 1952, JVN.

20. Duncan McDonald to James V. Neel, 22 August 1952, 6/T, JVN.

consistent than reports of location. He added that if they could not trust reports of symptoms, "then in view of the way the Japanese have of moving themselves around with respect to the hypocenter, none of our measures of radiation are worth a damn and this would seem to be a good time to close shop."[21]

There was some additional anecdotal evidence to support Neel and Schull. An early ABCC report prepared by Lowell Woodbury, Frank Connell, and Kenneth Noble included data that suggested some problems with the shielding model then accepted by the ABCC. Some survivors seemed to have been completely shielded at the moment of the bomb's detonation, but they had also suffered indisputable symptoms of radiation sickness.[22] This in itself suggested that shielding and distance should be rated as less important than radiation sickness in deciding how to classify a survivor.

The debate may appear trivial since it involved a relatively small number of survivors. But the numbers in all the exposed categories were relatively small. In the final published report in 1956, only 145 infants, of the 76,000 tracked in the five years of the study, were the offspring of parents who had both been heavily exposed (Neel and Schull 1956a, 45, 106–7). The heavily exposed category, then, was both the smallest and the most important, given the relatively low doses received by those who survived the blast and were also healthy enough to reproduce.

Because of an unrelated data-gathering decision the final analytical decision, to give symptoms precedence over distance, may have had an impact on the measurement of the radiation effect Neel and Schull considered most compelling: an effect on the offspring of heavily irradiated fathers and nonirradiated mothers. The data-gathering problem was the failure of the ABCC staff to take radiation histories directly from the exposed fathers included in the genetics study. The ABCC clerks and medical staff did not interview the majority of the fathers. Most information about paternal exposure came through wives at the time of the initial interview. The ABCC clerks taking radiation histories communicated with only about 22 percent of the fathers in the overall sample of pregnancy terminations in this early study. In many cases this com-

21. William J. Schull and James V. Neel to Duncan McDonald and Newton Morton, 22 August 1952, JVN.

22. Lowell Woodbury et al., "Preliminary Notes on Shielding and Distance as Factors of Exposure and Survival in the Nagasaki Atomic Explosion," drawer 23, U–W: Unpublished ABCC Reports, NAS. This report is undated but Woodbury left the ABCC in 1953, so presumably it was prepared no later than 1953. The copy in the NAS files includes a handwritten note on the front page, signed with the initials F. H. C. (possibly Frank H. Connell): "The following compilation of very preliminary and often inaccurate notes has been misinterpreted by almost everyone who has read it. This is not a manuscript intended for publication—it is a memo for Dr. [A. E.] Brandt [of the Atomic Energy Commission] written by Dr. Woodbury."

munication consisted of a postcard sent to fathers soliciting information about their location and experiences at the time of the bombing.[23] Generally, the study depended on each woman to report how far her husband had been from the hypocenter, what he had been doing, whether he had been inside, outside, shielded or in the open, and whether he had experienced any symptoms of radiation sickness in the days or weeks after the bombings. These were precisely the questions each woman was asked about herself. She was expected to remember her husband's experiences in as much detail as she remembered her own.

This may have been a less serious problem for husbands and wives who were together at the time of the bombings. But even in such cases, McDonald pointed out, it is one thing to remember one's own experiences during a crisis. "It is another matter to 'remember' the location and symptoms of another, even if that person is a spouse." In the cases of couples who met after the bombings, the wife's account would depend entirely on her husband's report, which may or may not have been complete or accurate.[24] Many survivors were reluctant to identify themselves as having been exposed to radiation when they sought a marriage partner. Fear of the genetic effects of radiation made survivors less desirable in the marriage market, and there is some evidence that arranged marriages were broken off if a survivor's status came to the attention of future in-laws (Ibuse 1969; Lifton 1968, 543–55; Kubo 1990). A male survivor might have good reasons to minimize his experiences, particularly if his bride were not herself a survivor.

McDonald had touched on one of the greatest difficulties with the radiation histories taken by the ABCC genetics project: the potential error in paternal radiation histories was high. If there were no pattern to the errors—that is, if the wives of exposed husbands were no more or less likely to recall the details of their spouse's experiences—then these errors might be of relatively little importance. As one ABCC staff physician noted in a report assessing the problem of different sources for radiation history, "the accuracy of the husband's symptom-history must surely vary depending on whether the husband usually gives it or whether the wife usually gives it."[25] When husbands were interviewed, as they were (belatedly) in the Kitamura early abortion study, the proportion of husbands reporting symptoms of radiation sickness was much higher than when, as in the overall genetics program, radiation histories were

23. Dr. Renwick, "Memorandum for the Record," 14 August 1952, 5/QCC, Kitamura Program, JVN.

24. Duncan McDonald to James V. Neel, 22 August 1952, 6/T, JVN.

25. Dr. Renwick, "Memorandum for the Record, 14 August 1952, 5/QCC, JVN.

taken only from the wives.[26] Since symptoms became the basis for the inclusion of any parent in the heavily exposed category (5), the failure of wives to report the husband's symptoms could result in a systematic underestimation of paternal exposure levels.

Neel and Schull stressed paternal effect as the only unequivocal sign of radiation damage. For this reason, marriages between an un-exposed woman and a heavily exposed man were important to their study. When the mother had been heavily irradiated, Neel and Schull commonly attributed abnormal births to the somatic effects of irradiation (the damaged maternal environment). A heavily irradiated woman who consistently bore sickly offspring, they argued, could be manifesting signs of somatic damage to her reproductive organs, rather than genetic damage. A heavily irradiated father's contribution to his offspring, however, was entirely genetic. A pattern of malformation in the offspring of heavily irradiated males (married, of course, to nonirradiated females) would provide truly convincing evidence of genetic change.[27]

Unfortunately, these were precisely the unions in which the father's radiation history, by McDonald's calculations, could be expected to be the least accurate and detailed—and least likely, significantly, to include information about symptoms. Wives who had themselves been irradiated, McDonald suspected, were more likely to know the details (symptoms?) of their husband's radiation experience, because a husband could be more forthcoming with a wife who was herself a survivor. Conversely, wives who had not been irradiated, but who were married to men who had been, were less likely to know details or symptoms of their husband's experiences.[28] Without information about symptoms, heavily irradiated husbands married to nonirradiated wives would be erroneously placed in lower exposure categories. This could result in a consistent tendency to underestimate the radiation exposure levels of these fathers.

In itself, this set of decisions might lead the investigators to overestimate genetic effects, since they would assume that a given effect was the product of a lower dose of radiation than it was. At the same time, however, the inclusion of the "lightly exposed" category 2 of survivors who had in fact been heavily exposed could also obscure a real genetic effect. This was because Neel, Schull, McDonald, and Morton chose to emphasize the comparison of heavily exposed (categories 4 and 5) with lightly exposed survivors (2), rather than

26. Duncan McDonald to Research Committee, 26 August 1952, drawer 3, ABCC Research Committee 1951 and 1952, NAS.

27. This argument is elaborated in Neel and Schull (1956a).

28. Duncan McDonald to James V. Neel, 22 August 1952, JVN.

with a control population composed entirely of un-exposed survivors (1). This simplified the problem of confounding variables but created problems of its own.

CONFOUNDING VARIABLES

Throughout the summer of 1952, the scientists struggled to decide how to compare their data. The confounding variables were numerous and perplexing. Many of the biological indicators the ABCC was tracking were believed to be affected by diet, maternal age, environmental teratogens, consanguinity (cousin-marriage), economic status, syphilis rates, and urban versus rural background. Schull viewed the fundamental question in June 1952 as "What is the effect of [exposure] on indicators, after removing the effects of the control variables?"[29] Neel and Schull sought the most economical way of removing these effects while retaining as much of their data as possible. There were two fundamental approaches that could be taken to deal with these variables. They could try to set up a regression analysis that would reflect the impact of these variables on pregnancy outcome, so that maternal age effects, for example, would not be confounded with radiation effects. Or they could dump the most problematic data until the differences in the remaining data were within acceptable limits, so that these differences could be more or less ignored in the analysis.

At first Neel, Schull, Morton, and McDonald suspected that time of conception might also be an important variable and that they would have to analyze the data by year to accommodate time trends.[30] They recognized that lumping together the data from all five years (1948–1952) would obscure changes through time. If there were unknown biological factors at work (such as repair mechanisms or selective elimination of germinal cells containing mutations), the proportion of abnormal births to survivor parents might fall over the years, in which case it would be a serious error to combine all the years.[31]

There was definite evidence that during the first year of collecting data in Nagasaki, the rate of late termination stillbirths was high. McDonald attributed this, however, to the mechanics of starting up the data collection program, rather than to any biological phenomenon. Women registered with the city and the ABCC when they were five months pregnant. Thus, for the first three or

---

29. William J. Schull to Duncan McDonald, 17 June 1952, 6/Q, JVN.

30. James V. Neel to Duncan McDonald, 23 June 1952, 6/Q, JVN.

31. Ibid.

four months of the program, the only reports of pregnancy terminations would have come from women who experienced a stillbirth in their sixth to ninth month. Women giving birth to full-term infants in the same period would not be included in the study, since they would have registered with the city *before* the ABCC study began. "This increases, of course, the stillbirth rate, and presumably changes the sex ratio and malformation rate," McDonald said, and therefore "anything dealing with rates should drop this year entirely." [32]

There were other possible factors involved in time trends. Neel noted in 1952 that there was definitely a reduction in the proportion of irradiated parents each year—survivors were moving out of Hiroshima and Nagasaki, while residents from other parts of Japan were moving in at high rates. In addition, the improving economic situation in postwar Japan affected nutrition, so that infants born in 1951, for example, could be expected to have a higher average birth weight than infants born in 1948. He doubted, however, that the accuracy of the observations had changed from year to year. Since birth weight was, they agreed, "a very poor criterion anyhow, and one on which we will lean more lightly than any other," he suggested pooling all the years in an effort to streamline the preliminary analysis. Confronted with so many complicated problems of analysis, Neel said the comparability of the data collected in different years now seemed, "of our various perplexing issues . . . one of the smaller." [33]

They had no choice but to disregard syphilis rates and economic status as variables, though both could affect rates of malformation, stillbirth, and neonatal death, and economic status (here equated with diet) had an influence on birth weight. The ABCC had collected data on syphilis only for the 10 to 30 percent of cases involved in the nine-month follow-up program. And the data on economic status were subjective: the Japanese physicians were asked to simply look around at the time of the home visit and to assess for themselves whether the family was very poor, poor, average, well-to-do, or rich (Neel and Schull 1956a, 61). These crude classifications did not suggest dramatic differences in economic status among the radiation exposure groups, that is, heavily exposed survivors did not appear to be more likely to be "very poor." [34] As in the case of radiation exposure level, economic status ceased to be an issue

32. Duncan McDonald to James V. Neel, 1 August 1952, 6/T, JVN.

33. James V. Neel to Duncan McDonald and Newton Morton, 12 July 1952, 6/Q, JVN.

34. Later analyses of the economic status of survivors indicated that those who were upper middle class before the bombings did not regain their status until 1966. Those who had been lower middle class before the bombings sunk lower, with declining income and living conditions. "Unskilled workers and day laborers, even those who had suffered little or no direct A-bomb damage, generally experienced great difficulty in recovering and sustaining their means of livelihood" (Committee for Compilation 1981, 431–33).

because the data were not available to meaningfully assess it. Similarly, syphilis rates were not considered in the analysis because they could not be assessed for 70 to 90 percent of the data. These variables were therefore, like year of conception, not compensated for in either the initial (1952–53) or final (1956) analysis.

Differences in parental age and parity (number of pregnancies) and in rates of cousin-marriage also complicated the analysis. Recent studies had suggested that the malformation rate increased with parental age. For a variety of complex social and historical reasons, irradiated parents were on average older than control parents. Schull hypothesized that many older survivors were "starting over" late (in their 30s), having lost children in the bombings. Among the controls, most parents were under age thirty-five; among the exposed, many persons over thirty-five continued to have children. This age factor was also later used to explain the average larger size of survivors' newborns: birth weight increases with maternal age. This could lead to an age-effect increase that would suggest a spurious (positive) genetic effect. The idea that survivors' offspring might be hardier or more fit as a consequence of radiation exposure did occasionally surface in both the internal discussions and the popular press. Neel and Schull explored the possibility of "radiation selection." Could the larger babies of heavily exposed survivors mean that the more "fit" individuals, of all those present in the two cities, survived the bombings and were able to reproduce?[35]

Conversely, rates of marriage to close relatives were lower among the exposed groups than among the control population. This was primarily due to the exposed sample in Nagasaki. Although Nagasaki as a whole had a higher rate of consanguineous marriage than Hiroshima, the hypocenter in Nagasaki was near Japan's largest Catholic church, Urakami Cathedral, and the heavily exposed group was disproportionately Christian. Catholics were religiously discouraged from cousin-marriage. Among Buddhist families, cousin-marriages were common and even encouraged, with some remote Japanese villages recording rates of cousin-marriage as high as 20 percent (Schull 1953). Among the controls in general, then, cousin-marriage was more common than among the exposed in general. This difference could lead to the dilution of a real genetic effect, since the controls could be expected to have a

---

35. After Schull discerned a significant increase in birth weight, not present at nine months, for both cities in the heavily exposed categories, Neel wondered "if by chance because of a 'radiation selection' the (4,5)'s [heavily exposed persons] are larger than average." James V. Neel to Duncan McDonald, 3 July 1953, 6/R, JVN. On the popular press, see my chapter 8. On Alice Stewart's conviction that the populations at Hiroshima and Nagasaki were not representative because of radiation selection, see Stewart (1982).

higher rate of malformation due to inbreeding than other populations, and this higher rate might conceal a higher rate due to radiation effects in the exposed.

In August 1952, Neel and Schull began thinking through how to accommodate these differences in the data sets. Clearly some data would have to be eliminated, but it was not clear which data sets were most expendable. Neel suggested that the "messiest" indicators—birth weight and nine-month anthropometrics—be left out in this first rough analysis: "The 'reference points' which are of the least value [birth weight and anthropometrics] are at the same time the messiest to handle, since our other reference points are essentially a 'yes' or 'no' type of category." He and Schull both felt "that there is something to be said for disregarding these reference points in the five-year analysis upon which we will base our decision to continue." In other words, Neel said, they might concentrate on early spontaneous termination, sex, malformation at birth, neonatal mortality, and malformation at nine months and "let the other slide temporarily, since we would probably not be greatly influenced by whatever we found in that respect."[36]

Dealing with maternal age and consanguinity appeared to pose more serious problems. But McDonald and Morton proposed a solution that would quickly solve the problem of both cousin-marriages and over-age mothers. "Why not eliminate consanguinity and age effects by eliminating all cases of consanguineous marriages and of mothers over a certain age" from the analysis? McDonald acknowledged that it sounded "drastic," but felt it would dispose of the problem "clearly and neatly." One clear disadvantage to such an approach was that it would cut deeply into the "both exposed" category. A preliminary analysis carried out by Morton showed that parents who were both exposed to the bomb were generally older than unexposed parents. Less than one-third of the parents included in the ABCC genetic study were exposed at all, and only 3.22 percent of all registered parents in Nagasaki and 7.21 percent in Hiroshima fell into the "heavily exposed" categories 4 and 5 (Neel and Schull 1956a, table 2.1, 19). Eliminating mothers over age thirty-five would shrink this group even further, but, McDonald said, "if age is a major factor, then it has to be eliminated or corrected for."

McDonald estimated that the elimination of both consanguineous marriage and the over-thirty-five age group would result in a loss of about 17 percent of the total data. "A large loss, but possibly worth it, considering the immense simplification achieved."[37] They knew cousin-marriage increased the frequency of "untoward pregnancy outcomes." Morton's own reviews of age

---

36. James V. Neel to Duncan McDonald, 21 July 1952, 6/Q, JVN.
37. Duncan McDonald to James V. Neel, 1 August 1952. JVN.

and parity showed a sharp increase in malformations when mothers were over age thirty-five, but uniformity among younger age groups. Cutting out the older mothers would also cut down the high parities. (Parity refers to the number of pregnancies.) A newborn with two older siblings, and whose mother had experienced one stillbirth, would have a parity of four, that is, he would be classified as the outcome of the fourth pregnancy. Since mothers who have borne many children are generally older than those who have borne only one or two, high parity and age were related complicating factors. By cutting out all mothers over age thirty-five, the ABCC investigators could solve two problems at once. By cutting out all cousin-marriages, they could avoid having to compensate for the effects of consanguinity in their initial analysis.

This proposal, outlined in a letter from McDonald and Morton to Neel and Schull, induced in Neel an "attack of acute intellectual indigestion." It took him a while to reach "the point where I trust I can write a coherent letter." Neel and Schull had proposed setting up a sort that would allow them to compare malformation and other indicators within comparable age and consanguinity subgroups. That is, the offspring of consanguineous marriages would be compared for exposed and control, and the offspring of marriages involving mothers over age thirty-five would be compared for exposed and control, and these selected groups of offspring would also be compared to the total sample.[38] This plan had some disadvantages, since it threatened to subdivide the data so minutely that some points of comparison would be impossible. If no neonatal deaths had been recorded, for example, among mothers over age thirty-five classified in radiation category 3, then that cell would contain a zero. McDonald felt that dumping all mothers over age thirty-five and all consanguineous marriages would dispose of the problem of such small categories. "How to analyze such material?" he asked. "The data is subdivided to an analytically impossible degree."[39]

Neel, in his initial response to McDonald's proposal, explored the question of low-value cells. He agreed that maternal age and consanguinity had to be accommodated in some way in the analysis. But discarding the data for all mothers over thirty-five and all cousin-marriages would prevent them from extracting a good deal of material of biological interest from the tabulations, he said. "What you propose . . . would not give us the data at all." The empty cells did not worry him. He recognized that "for certain exposure categories, and possibly for the majority of them, the numbers involved in some of the cells would be entirely too small for a chi-square type analysis. But it seems to me

38. James V. Neel to Duncan McDonald, 21 July 1952, 6/Q, JVN.

39. Duncan McDonald to James V. Neel, 1 August 1952, 6/T, JVN.

it is better to have the data and then disregard it if forced to, than not to get it and, quite possibly, end up by wishing one had gotten it." He added that if they followed Morton's proposal "we will not extract from our data a good deal of the material pertaining to consanguinity and age effects which have interest quite apart from the question of atomic bomb effects."[40]

Within ten days, however, Neel and Schull decided that McDonald's plan should be tried. This was a change so precipitous that they sent two cablegrams to Japan immediately telling McDonald to do as he had proposed and eliminate from the analysis the older mothers and cousin-marriages. This change did not reflect any change in their view of what the ideal scientific approach would be. It was, rather, a practical compromise: McDonald's plan was more likely to provide the quick answer needed to facilitate formal approval to continue the genetics study. It was an analytical decision clearly shaped by the immediate need to produce results that would convince the NAS and the AEC that the project should be continued.

Cutting out older mothers and consanguineous marriages would not do for the final analysis, Schull said in a letter that followed the cables. But if the plan to cut out consanguineous marriages and older mothers would produce unequivocal (and rapid) results, it would help determine whether the program should continue. It should be "borne in mind," however, that by eliminating mothers thirty-five and older they were merely "truncating a continuous distribution at a point where maternal age effect would be minimized *if* the remaining age groups are comparable." If they were not comparable—and a later analysis suggested they were not—the age problem remained.[41] He added that the truncated analysis would not even do for the preliminary analysis if it produced equivocal results. "It is largely because of the urgency of deciding whether or not genetics should be terminated that the 'go ahead' was given. I think then that we are all agreed that this is a 'pilot study.' "[42] Neel, in a letter accompanying Schull's, reiterated that this was only a "pilot study" and said that he hoped they would be able to make allowance for consanguinity and maternal age in the final study (which they did in the 1956 publication). "This was not a happy decision on our part despite the obvious simplification of tabulating and analyzing the data which such a decision produces."[43]

40. James V. Neel to Duncan McDonald and Newton Morton, 12 August 1952, 6/Q JVN.

41. William J. Schull and James V. Neel to Duncan McDonald and Newton Morton, 22 August 1952, 6/Q JVN.

42. William J. Schull to Duncan McDonald and Newton Morton, 22 August 1952, 6/Q, JVN.

43. William J. Schull and James V. Neel to Duncan McDonald and Newton Morton, 22 August 1952, 6/Q, JVN.

The need to produce results made sense of the decision to eliminate 17 percent of their data and a disproportionate percentage of their data on the offspring of two heavily exposed parents. In a way the two problematic variables dispensed with in this decision complemented each other. The relatively higher rate of older mothers among the exposed might create spurious genetic effects. The relatively higher rate of consanguineous marriages among the controls could obscure real genetic effects. By dumping both sets of data, they eliminated a whole range of problems of comparability.

CONSTRICTING THE DATA BASE

The decisions made in the 1952–53 analysis, including the decision to exclude mothers over age thirty-five and consanguineous marriages, had both practical and theoretical justification. They were shaped by the need to produce results that would be useful in assessing whether the genetics study should be continued. The four geneticists involved hoped that their analysis would provoke the Committee on Atomic Casualties to support continuing the genetics project. Instead, participants in the July 1953 genetics conference in Ann Arbor made the decision to cease collecting morphological data—basically terminating the ABCC genetics project.[44] At this meeting, Neel and Schull reviewed their findings, presenting both their own analysis and an independent, somewhat different analysis carried out by Newton Morton and Duncan McDonald in Hiroshima. They concluded that the results indicated that the pessimism about the program expressed at the first genetics conference, in June 1947, was justified. Some obvious problems included the small number of heavily irradiated parents, the practical difficulties of the study, the ambiguity in mutation indicators, and the difficult statistical problems. They could detect no significant differences among the offspring of heavily and lightly exposed survivors for gross malformation, spontaneous abortion, neonatal death, or anthropometrics at age nine months. Significant differences were possible for type of termination, sex ratio, and birth weight. The birth-weight data were contrary to the genetic hypothesis, since the offspring of exposed parents tended to be larger and heavier.

---

44. George Beadle, Don Charles, C. C. Craig, L. H. Synder, and Curt Stern were the official members of this conference. Also attending were Atomic Energy Commission officials A. E. Brandt, H. H. Plough, B. R. Nebel, and E. L. Green; Philip Owen of the National Research Council; John J. Morton of the ABCC; and Lee Dice and other members and associates of the Institute of Human Biology of the University of Michigan. G. W. Beadle, D. R. Charles, C. C. Craig, L. H. Snyder, and Curt Stern, "Statement of the Conference on Genetics to the Committee on Atomic Casualties," 11 July 1953, 6/R, JVN.

The type of termination data (stillbirth versus live birth) was significant only for maternal irradiation, but, as Neel observed, higher rates of stillbirth in irradiated mothers might be a sign of somatic damage to the mother rather than genetic damage to the fetus. Only the sex-ratio data seemed to hold up, though this data too could be interpreted in various ways.[45]

These results were of some interest—but they did not seem to justify the continued expense and effort of full-scale data collection at Hiroshima and Nagasaki. "It is now apparent that a large majority of the children of exposed parents has already been born so that the reduction in present statistical errors afforded by presumed 1953–1957 births would be small." The genetics conference therefore "unanimously recommended that collection of field data for the Genetics Program be terminated by June 1954 unless, on the basis of new findings, a recommendation for continuance be made by the conference on Genetics." The conference favored in principal a "future study of the fertility of the children of exposed parents" and the possible continuation of the collection of sex-ratio data. But the vote effectively terminated the laborious and massive system of infant examination at Hiroshima and Nagasaki.[46] This system was never revived, and though the ABCC and its successor agency, the Radiation Effects Research Foundation, have since carried out a wide range of other studies to assess the genetic effects of radiation, including a biochemical genetics study and a study of somatic mutations in survivors' children, the program of examining newborn infants for morphological anomalies ceased in January 1954.

Virtually all scientists dump data and in a study of this type some dumping was inevitable and necessary. What is significant is not that the data were dumped, but the relative importance attached to confounding variables in the decisions about what data to eliminate. Radiation exposure was merely one of many variables. As the subject of the study it was by definition not a confounding variable. The purpose of the study was to decipher the effects of radiation on certain indicators. But the effects of many of the confounding variables in this study were no better understood than were the effects of radiation on the indicators selected for analysis. Some parameters known to affect the data set (such as syphilis) were ignored because the data were not available. Others with unknown effects (rural background) were granted importance in the partitioning of the data.

---

45. The proportion of males born to control mothers was 51.92 percent, while the proportion for heavily irradiated parents was 50.16 percent. G. W. Beadle, D. R. Charles, C. C. Craig, H. L. Snyder, Curt Stern, "Statement of the Conference on Genetics to the Committee on Atomic Casualties," 11 July 1953, 6/R, JVN, 2.

46. Ibid., 3.

Some data dumping seemed to "simplify" the analysis without threatening its validity. Yet at each step in this analysis, reasonable arguments could be made (and sometimes were) for eliminating data in a different way. For example, the age and consanguinity effects could have been controlled for by eliminating a selection of "control" parents under age thirty-five until the age proportions for exposed and controls were matched—this approach was briefly considered by Neel and Schull. The ABCC team might also have decided to compare only the births to heavily exposed parents, to a sample of unexposed parents matched for age and consanguinity; this had been one of the original suggestions explored at the first genetics conference in June 1946. In this way they could have retained their potentially most revealing data. They might also have tried to compensate for the higher expected rate of malformation in over-thirty-five mothers based on data from other populations indicating the rate of various malformations for these mothers, though they were generally reluctant to compare the Japanese populations to other ethnic populations for which such estimates had been made. It would be possible to argue that the most important thing was to get as much as possible from the small mass of heavily exposed data, particularly since radiation effects on heredity in these populations were expected to be relatively small.

In July 1953 Duncan McDonald suggested to Neel that he should purposely minimize radiation effects on the survivors at Hiroshima and Nagasaki in the preliminary report of the genetics project. McDonald felt that the report could not possibly "present clearly and understandably the full significance . . . of our findings" and could therefore be readily misinterpreted (particularly in the popular press). He said that Neel should "at all costs, distasteful though it be, deliberately minimize radiation effect indications." The birth-weight data should be "presented with the simple statement that no radiation effect has been demonstrated. The stillbirth effects should be discounted in every possible way. And the sex-ratio effect should not be unduly stressed, the alternate somatic cause or complication should be put forward, and it should be indicated that further work is being done in an attempt to find if this is a true effect and not a random variation such as occurs in any sampling process."[47]

Neel bristled at the suggestion. "I find myself at a loss, as I am sure would be the majority of geneticists, to understand why we must 'deliberately minimize radiation effect indications.'" They should "present the truth" to their scientific colleagues, Neel told McDonald, "not what you feel they should have."[48]

47. Duncan McDonald to James V. Neel, 31 July 1953, 6/V, JVN.

48. James V. Neel to Duncan McDonald, 10 August 1953, 6/T JVN.

McDonald, however, was acknowledging a certain kind of truth: their results would be simplified and manipulated by the various voices in the public debate.[49] McDonald suspected it would be safer to minimize genetic effects, since genetic effects were clearly dangerous in the public debate. But the public debate included both persons who sought to minimize radiation effects (such as some officials at the AEC) and those who sought to maximize them (such as critics of atmospheric weapons testing). While McDonald worried about the ABCC results being manipulated by those who would exaggerate genetic effects, Neel also worried about the results being manipulated by those who would minimize them. Neel was concerned that the Atomic Energy Commission would misrepresent the ABCC findings to the public, using them to declare that no genetic effects had occurred at Hiroshima and Nagasaki. He was "disturbed by the suggestion that the work might be interpreted as placing more emphasis on the failure to detect [genetic effects] and less on the failure to prove that there had not been an effect, which could not necessarily have been detected."[50] The problem of the public debate about radiation effects did not necessarily dictate a specific result, but it did mean that the results would receive a great deal of scrutiny, and it encouraged extreme caution, or conservatism, in all assessments of data.

The analysis as undertaken was widely supported by the broader scientific community. The statistical approach was open to negotiation, but fundamentally the choices made were acceptable to both statisticians and geneticists who reviewed them. "Our problem is one of sufficient complexity that it is doubtful whether any two competent statisticians would solve it precisely the same way," Neel observed, and this proved to the case.[51] Yet in general, and particularly at the joint meeting of the Genetics Society of America and the Biometrics Society, at Ames, Iowa, in July 1952, their plans were "reasonably well received," meaning that "no one expressed any particular dissatisfaction with the way we had approached our problem, recognizing it for the very messy problem that it is."[52] By 1956, when the results were formally presented in a monograph, the messiness of the data set and the analysis was further complicated by a political climate in which the ABCC genetics results seemed to have a significant bearing on the fate of humanity—and on presidential politics.

49. For the public debate on radiation effects, Robert Divine's *Blowing on the Wind: The Nuclear Test Ban Debate 1954–1960* (1978) is a helpful source.

50. Sterling Emerson (AEC geneticist) to the U.S. Atomic Energy Commission, Advisory Committee on Biology and Medicine, 26 November 1956, drawer 19, NAS.

51. James V. Neel to Duncan McDonald, 23 June 1952, 6/Q, JVN.

52. James V. Neel to Duncan McDonald, 21 July 1952, 6/Q, JVN.

CHAPTER ELEVEN

Publication Strategies

T he first publication of the ABCC genetics summary results appeared in the 6 November 1953 *Science*. The comprehensive presentation of the same data followed in a 1956 monograph published by the National Academy of Sciences as *The Effect of Exposure to the Atomic Bombs on Pregnancy Termination in Hiroshima and Nagasaki* (Neel and Schull 1956a).

When the ABCC genetics study began in Japan in 1946, the risks of significant radiation exposure seemed to be limited to a few unique populations: atomic bomb survivors (either at Hiroshima or Nagasaki or at future locations), workers in some industrial settings, and patients treated with radiation. By the time the results were published, in 1956, atmospheric weapons testing and the discovery of global fallout had expanded the risk to include virtually every living human being.

The March 1954 Bravo weapons tests, in which 236 Marshall Islanders and twenty-three Japanese fishermen suffered radiation sickness from "fall-out," kicked off a public panic in both Japan and the United States (Divine 1978, 6–8; see also Lapp 1958). The U.S. Congressional Joint Committee on Atomic Energy held its first public hearing in April 1955 (Divine 1978, 43–47). The United Nations International Conference on the Peaceful Uses of Atomic Energy met in Geneva in the summer of 1955.[1] Later the same year the United Nations set up its Scientific Committee on the Effects of Atomic Radiation, UNSCEAR (World Health Organization 1957). In July 1956, the National Academy of Sciences prepared the first of two "reports to the public" on the biological effects of radiation,[2] and the Medical Research Council of Great Britain, similarly, published a report on the effects of radiation on human heredity (Medical Research Council 1956).

Radioactive fallout also became an issue in the 1956 presidential campaign. In a speech in Los Angeles on 5 September 1956, Democratic candidate Adlai Stevenson called for a unilateral end to weapons testing on the grounds that weapons testing threatened the biological future of the species. Until this speech, the risks of fallout from weapons testing had not played an important role in the campaigns of either Stevenson or his Republican opponent, Dwight Eisenhower. After this speech, the issue was widely debated by the candidates and the press, and Stevenson successfully used it to attract public support, though this support dissipated in the mists of Cold War politics: Soviet endorsement of his plan led to public rejection in the United States (Divine 1978, 86–109).

Neel and Schull, preparing the final draft of their book manuscript in early 1956, felt the urgency surrounding the issue. In February 1956 Neel believed that there was "a great deal of pressure" on them both "to get the Japan material out just as soon as it can conveniently be arranged." He told one of his editors that "our preoccupation with the time element may seem a little foolish to you, coming as it does at the end of a ten-year haul. I can only repeat our conviction that for a variety of reasons this should be gotten out very shortly."[3] By July 1956 the final report was ready. Neel and Schull presented papers on their

---

1. H. J. Muller was scheduled to speak at the conference, but approval for his talk was denied by AEC officials when they learned that he planned to mention Hiroshima and Nagasaki. See Beatty (1987).

2. Members of the committee producing this report were Warren Weaver, George Beadle, James Crow, M. Demerec, G. Failla, H. Bentley Glass, Alexander Hollaender, Berwind P. Kaufmann, C. C. Little, James V. Neel, W. L. Russell, T. M. Sonneborn, A. H. Sturtevant, Shields Warren, and Sewall Wright (National Academy of Sciences, National Research Council 1956; National Academy of Sciences, Committee on the Biological Effects of Atomic Radiation 1960).

3. James V. Neel to A. G. Steinberg, 9 February 1956, 5/HB, Japan Publishing Business, JVN.

findings at the First International Congress of Human Genetics in Copenhagen in August 1956 and at the International Genetic Symposia in Tokyo in September 1956.[4] The final analysis was somewhat different from that presented in their preliminary publication. The most important difference was the disappearance of the sex-ratio effect suggested in the 1953 paper in *Science.*

This chapter examines the creation and form of the two texts, published in 1953 and 1956, that described the ABCC genetics results. I consider patterns of information inclusion and emphasis, suggesting how they reflected the authors' sensitivity to the public impact and institutional meaning of their work.[5] I explore how some of the crucial wording was negotiated by Neel, Schull, their coauthors, and their commentators at the ABCC.[6] I examine the reasons for the timing of the publications. And I show how the 1956 book, particularly, reflected the broader debate over radiation and atmospheric weapons tests in its conclusions and explanatory style.

The 1953 *Science* essay and the 1956 monograph were carefully worded explanations and justifications of the ABCC approach. For the public and the scientific community at large, these texts *were* the ABCC genetics study. In many ways, they remain its most important record. My analysis draws on the archival data to explore how this public presentation was assembled.

4. Sterling Emerson, "Genetic Effects of Radiation Exposure," report and memo, 26 November 1956, drawer 19, Genetics—1956–1957, NAS, 6.

5. Following Myers, I interpret these texts as part of a process through which knowledge claims are constructed (Myers 1990).

6. Though there were AEC officials who would have been happy to control all ABCC publications, this was not possible. On only one point did the AEC or its advisory committee members encourage Neel to change his wording and this was a change that left more room for genetic effects. That is, the AEC geneticist suggested that Neel restrict his claims regarding the level of genetic effects which were ruled out by the study. The exact sentence, which appeared in the 1956 book, said the results were "sufficient to exclude the possibility of 'large' genetic effects but . . . still compatible with the range of effects observed in mice and *Drosophila.*'" Neel had originally written that the results excluded "average" genetic effects. The use of the word "large" was a consequence of negotiations with AEC geneticist Sterling Emerson. Emerson felt that "average" did not have a "clear meaning," but his suggestion, by limiting the results to "large" genetic effects (and both these terms mean very little), restricted the scope of the Neel-Schull results. Emerson discussed this in a meeting of the AEC Advisory Committee on Biology and Medicine, "Minutes," 26 November 1956, box 3218, DBM-ACBM Minutes, RG 326, DOE, 147–48.

PRELIMINARY PUBLICATION

The genetics results made their first formal public appearance in a brief paper in the 6 November 1953 issue of *Science*.[7] The publication of this relatively mild-mannered, cautious paper was preceded by months of debate about its purpose and form. The paper was not based on a consensus at the ABCC that the time had come to tell the world about the genetics results. Rather, the results were published in the United States on 6 November because an ABCC scientist in Hiroshima, Duncan McDonald, had agreed to present preliminary results at a Japanese scientific meeting on 7 November. Such a unilateral announcement of preliminary results—an announcement only in Japan—would presumably have gratified the Japanese scientific community, but some members of the Committee on Atomic Casualties feared that it would also attract significant press coverage, and that the message would be garbled in traveling from Japan to the United States. Others felt that scientific courtesy demanded that the results be published simultaneously in both Japan and the United States.[8] Such objections prevailed: the *Science* paper, then, was published in order to balance the presentation of results in Japan.

In early summer, McDonald agreed to present the findings to the Japanese Society of Genetics in Tokyo that fall. Meanwhile, the group of scientific advisers to the genetics project decided that most of the collection of data on newborns should cease (see my chapter 9). Neel felt that, while McDonald's presentation was "a fine thing to do" and "in line with our announced policy" of keeping the Japanese scientific community informed, it might have negative consequences. "It might be well to have a more or less simultaneously timed article in the United States," he said, since Japanese reports would be picked up by the international press.[9]

It was also important that McDonald's presentation in Tokyo be perfectly consistent in tone and content with whatever was published in the United States. McDonald's tendency to dismiss differences between the exposed and the control groups as a consequence of forces other than radiation was also a problem. If McDonald told the Japanese scientists that no genetic effects had occurred, while a report to the American scientific community said they had,

7. There had been one earlier report, prepared by John C. Bugher of the Atomic Energy Commission and published in *Nucleonics* (1952, 18), which suggested a significantly higher rate of major malformation among the offspring of exposed. This finding was not consistent with the 1953 report by Neel and Schull.

8. John Beatty also explores the sequence of events leading up to this 1953 publication (Beatty 1991, 303–304), suggesting that Neel feared the public relations consequences of simultaneous, nonidentical publication in Japan and the United States.

9. James V. Neel to Grant Taylor, 14 July 1953, 6/R, JVN.

the disparity would be noticed. Neel and Schull assessed the data as suggesting at least the possibility that genetic effects could be demonstrated for sex ratio and stillbirth. McDonald believed that such differences as appeared in the data were the consequence of other differences in the exposed and control groups. Simultaneous and conflicting messages might lead Japanese critics to conclude that the ABCC was trying to mislead the Japanese scientific community, while reporting the truth in the United States.

Neel noted that, "strangely enough," there were also those who felt that "American scientists were at least entitled to a simultaneous presentation with Japanese scientists." [10] He was not at all sure that the time had come for a full publication—"I would be perfectly happy to call this entire matter of publication off until such time as we carry through certain secondary tabulations which may serve to show definitely whether the stillbirth and birthweight finding are each a parity effect"—but he told McDonald, "it would be a rather pleasant experience for you to present this material before the Japanese Genetics Society, and I would certainly not want to upset your plans at this time." [11] Neel personally did not expect McDonald's report to have much "news value," but George Beadle and Curt Stern, serving on the Committee on Atomic Casualties, felt that coverage would be extensive and "feared that after it had passed over several rewrite desks on its way from Japan, the end result might be pretty garbled." [12] Neel suggested publication in *Science,* to coincide or slightly precede the presentation in Japan. "The press will probably one way or another manage to run a story, and we really have little alternative but to do the best we can to assure an accurate presentation the first time, since as we are all aware, the denial is ever so slow in catching up." [13]

ABCC director Grant Taylor, however, objected to simultaneous publication in the United States. His view apparently reflected a strong current of feeling among the American staff in Japan. There were obvious reasons to favor a formal announcement of results in Japan only. Some of the Japanese resentment about lack of access to ABCC data might be assuaged, and the announcement could be interpreted as recognition of the significant contributions of the Japanese, both as subjects of research and collaborators with the ABCC, in the study. But Taylor's articulated objections to the simultaneous publication seemed to derive rather from his feeling that Neel, who was in Ann Arbor, Michigan, was meddling. He did not specifically oppose publication in

10. James V. Neel to Duncan McDonald, 10 August 1953, 6/R, JVN.

11. James V. Neel to Duncan McDonald, 31 August 1953, 6/R, JVN.

12. James V. Neel to Grant Taylor, 14 July 1953, 6/R, JVN.

13. Ibid.

Science, but he interpreted it as one more example of Neel overmanaging the genetics study, which he believed should be controlled by those on the scene in Japan. ABCC staff in Hiroshima knew how best to handle the situation, he said, and Neel should defer to their judgment. He particularly questioned Neel's plans to handle the statistical analysis for the final monograph in Ann Arbor. Taylor felt that the analysis should be handled by the staff in Hiroshima. "We have a competent statistics department, and I feel that during the last year we have learned the difficulty of trying to keep the Ann Arbor group in touch with these data." He added that the "Japanese who have worked intimately with these data are an asset which should be retained. Labor such as this is cheap in Japan." [14]

In August, Taylor addressed his objections not to Neel himself, but to the director of the AEC's Division of Biology and Medicine, John Bugher, then also serving on the NRC's Committee on Atomic Casualties. Bugher agreed with Taylor that the "fundamental statistical work should be done at Hiroshima." He added that he would not approve publication in *Science* without the specific endorsement of Taylor, and in any case, "on statistical grounds alone I would not agree to publication of any of this at the present time." [15]

Neel and Schull meanwhile worked up a draft for possible publication, based on their data analysis over the previous eighteen months (see my chapter 10), and sent it to the group in Hiroshima for review in July 1953. This draft opened by stating that "for the past seven years studies have been underway in Hiroshima and Nagasaki, Japan, designed to provide answers" to two questions: First, were there any differences in the children of exposed and control groups? And second, if differences did exist, how could they be interpreted? The draft explained the radiation classification of parents, and stated that the analysis it presented would be limited to mothers less than thirty-five years old—about 84 percent of the Hiroshima and 77 percent of the Nagasaki registrations. In exploring the question of the larger birth weight of the offspring of survivors, Neel and Schull raised the possibility of somatic selection. The findings on birth weight were "puzzling," they said, but there were two possible explanations: First, parity is significantly related to birth weight and parity differences in the exposed and the control groups persisted despite the decision to eliminate all mothers more than thirty-five years old. Second, Neel and Schull said, heavily irradiated survivors "may represent a somatically selected

14. Grant Taylor to John C. Bugher, 28 July 1953, drawer 6, Miscellaneous Correspondence outside Japan 1952–1960, NAS.

15. John C. Bugher to Grant Taylor, 4 August 1953, drawer 6, Miscellaneous Correspondence outside Japan 1952–1960, NAS.

group of Japanese, larger and heavier than the average, with this fact reflected in their infants." They pointed out that their approach permitted "the detection of only a fraction of the total genetic effect of exposure to an atomic bomb" but taking the results "at face value they can be interpreted as indicating that the sensitivity of human genes to irradiation is of the same order of magnitude as *Drosophila* or mouse genes." In the cover letter sent with this draft, Neel asked McDonald to respond quickly. "Time is of the essence. Could you cable me if there is approval or indicate that a letter follows if there are strong objections?"[16] As Neel learned, there were indeed strong objections to this draft.

The ABCC staff in Japan reacted to the Neel-Schull draft with general disapproval on a wide range of points both stylistic and substantive. Lowell Woodbury, in the statistics department, felt that the identifying organization for Neel and Schull should be the ABCC, rather than the University of Michigan (this suggestion was followed in the published version). Earle Reynolds (later a nuclear activist) felt there were too many acknowledgments and that "if acknowledgments are to be made in the paper, only the Japanese midwives, physicians and patients should be included." (Neither Reynolds nor any of the other readers noticed that the ABCC's Japanese liaison agency, the Japanese NIH, was not thanked in the paper). Reynolds further recommended that instead of a full-fledged article, the submission to *Science* should be a thousand-word summary of McDonald's planned remarks before the Genetics Society of Japan. "It would be stated that the purpose of such a note is to prevent misunderstandings due to difficulties of communication."[17]

But the most aggressive response came from McDonald, who wrote a seven-page detailed attack on the language, reasoning, and assumptions of the proposed publication.[18] His concern was grounded partly in his conviction that "this paper is going to be the one which is quoted, not the definitive monograph to follow. We cannot be too careful now."[19] He said that the draft prepared by Neel and Schull gave too much credibility to the sex-ratio effects. "It must be stressed that there is considerable doubt about a *radiation* interpretation of this difference. . . . There is so much doubt about this finding that we must hedge in every way [and] stress *strongly* that this may be a maternal somatic effect and not a genetic effect" (emphases in original). He criticized the wording

16. James V. Neel to Duncan McDonald, draft manuscript, 21 July 1953, notebook 5, Genetics, Correspondence II, WJS, AA.

17. "Research Committee Meeting—Discussion of Genetics Paper," 31 July 1953, drawer 3, ABCC Research Committee 1952, NAS.

18. Duncan McDonald to James V. Neel, 31 July 1953, drawer 3, ABCC Reearch Committee 1952, NAS.

19. Duncan McDonald to James V. Neel and William J. Schull, 31 July 1953, 6/V, JVN.

describing malformations, noting that the text said "no clear tendency" toward malformations was present, but "this indicates that a tendency has been discovered, somewhat mixed. Change this to '. . . no such tendency.' " He said Neel and Schull's conclusions on birth weight "contravene[d]" the evidence. "There is no exposure effect on birthweight. Your two tables contradict one another." The numbers in their tables were "hopelessly small," and "the fact that your second table, tripling the number in the exposed group, changes the difference from a significant increase to a non-significant decrease must, I think, disturb any reader who looks at the tables at all carefully. You do not mention this rather serious discrepancy in your discussion." He also criticized Neel and Schull for suggesting that the data compiled at Hiroshima and Nagasaki were consistent with results from mouse and *Drosophila* studies. "In what way are our findings consistent with anything, and what is the evidence? As far as our data indicate there may be *no* genetic effect of radiation in humans. . . . this is not true of *Drosophila* or mouse. If you (and Dr. [Curt] Stern) must bring in this extraneous group of words ["consistent with what is known of the radiation genetics of a wide variety of plant and animal material, including *Drosophila* and mice"], you must at the very most say that our data, given our guesses at radiation dosages, would seem to indicate that radiation effects may well be no greater in humans than in experimental organisms, and add that, also, as far as the data go, there may be no radiation effect at all." He closed this letter by summarizing the objections of other staff members in Japan and suggesting that the entire staff considered the paper "diffuse and hasty."[20]

Neel interpreted McDonald's letter as a "violent and, in my considered opinion, irresponsible diatribe," and responded point for point to McDonald's objections.[21] He called McDonald's questions about the simplification of groups for comparison "completely specious." McDonald's questions about the large number of controls, which were being compared to a significantly smaller group of exposed, he dismissed. "It just so happens that there *are* more controls than there are heavily irradiated. This is a feature of the data which neither I nor anyone else can change, although you may be able to find some way to overlook it, which is not at the moment clear to me." To McDonald's observations on the comparison to mouse and *Drosophila,* Neel replied that the issue had to be broached. "Our statement is to the effect that taking the data at face value, they are consistent with what is known of the irradiation genetics of other animals. They do not indicate that human genes are either

20. Duncan McDonald to James V. Neel, 31 July 1953, drawer 3, ABCC Research Committee 1952, NAS.

21. James V. Neel to Duncan McDonald, 10 August 1953, 6/R, JVN.

more or less sensitive than, for example, mouse genes. It is felt important by a number of geneticists whose judgment I have come to respect that this aspect of the question be mentioned in the initial presentation." He closed his assessment of McDonald's remarks by stating that he was "completely baffled as to the motivation behind this attempt to discredit the *Science* article so completely." However, he suspected ulterior motives: "You must be well aware of the consequences of your action. I have no choice at the moment but to circulate this answer rather widely since your action has undoubtedly raised questions in certain quarters. At the very least, you have rendered it difficult to arrange a simultaneous release in this country with the presentation in Japan, a consequence which may not come as a complete surprise to you." Neel thus accused McDonald of attempting to sabotage the publication in *Science* so that the results would be announced only in Japan.[22] On the same day he sent his response to McDonald, Neel also appealed to Bugher. He explained that while Morton and McDonald retained some reservations about the proposed article, these could be readily resolved. He asked for permission to pursue publication prior to the 13 September meeting of the Committee on Atomic Casualties, in order to assure publication in time for the November presentation in Japan. Bugher, perhaps more sympathetic to Taylor and McDonald than to Neel, agreed that Neel could seek publication before he appealed to the entire committee at the September meeting, provided he addressed the questions raised by the group in Hiroshima.[23]

Neel felt that the "crux of the difference" was in McDonald's view of what should be made of the differences between exposed and control groups that did appear in the data. "At the moment we observe certain small differences between the controls and the lightly irradiated on one hand, and the more heavily irradiated on the other hand," Neel observed. McDonald, he said, was "completely convinced that these are not due to irradiation as such, but rather due to differences in the two groups under comparison. He is so convinced of this fact that he would like to present the data in such a way that these differences do not appear." This was the suggestion that had troubled Neel throughout his discussions with McDonald, for while Neel did not want to exaggerate effects, neither did he want to manipulate the data to make differences between exposed and controls disappear. Neel and McDonald agreed generally that any differences should be interpreted "conservatively," that is, as not necessarily indicating radiation effects. But they disagreed about how this problem should

22. Ibid.

23. John C. Bugher, director, Division of Biology and Medicine, "Conversation with James V. Neel," memo to files, 10 August 1953, drawer 19, Genetics—1949–1953 #3a, NAS.

be handled for public presentation. Neel said McDonald might be correct—the differences in heavily and lightly exposed groups might be the consequence of factors other than radiation—but said that "until we are sure where we stand it is better to present the whole story."[24]

McDonald's response to Neel's counterattack was to both apologize and insist that he was still right. "I am very sorry indeed that my letter was so written that it could be misinterpreted. It was based entirely on an honest and fundamental criticism of your approach to the data." This fundamental criticism—which was perhaps obscured, McDonald said, by the "mixed bag" of objections in the earlier letter—was that the control population was not an adequate group for comparison to the survivors. Indeed, the controls "may well never be a suitable control" and "certainly cannot be used except for major malformation and sex ratio." The control mothers were significantly younger than the exposed mothers, and any differences in their offspring could therefore be age effects. In addition, the controls were more likely to be of rural origin or to be repatriates from Manchuria and Korea. "The effect of this on our indicators is unknown but may well not be small."[25] McDonald included with his formal letter to Neel a personal letter dated the same day. "What in the world is up with you?" he asked. "I consider it rather unfair that you should impute that I or others here have ulterior motives in our criticism. Do you mean that I want to publish first in Japan with my name on the paper? . . . As far as I know, I am the one disinterested person amongst the main people concerned. I do not care if my name appears in print. I have no reputation to make or maintain in human genetics. But perhaps, and I do not blame you, you have read my character differently."[26]

Neel's response to this letter was more tempered. He said he had been "genuinely surprised" when "you threw the book at me" and said a "good deal" of McDonald's criticism was "unjustified." But he agreed that "it would be too bad if after all the trials and tribulations through which Genetics has presented a relatively united front, a serious rift should develop at this time." He noted that there were three possible differences between the offspring of the heavily irradiated survivors and controls, differences that all parties agreed existed. These were sex ratio (probably the most significant), frequency of stillbirths, and birth weight. If all the unexposed in the ABCC's category 1 (not present in the cities at time of bombing) were eliminated, Neel said, these differences

24. James V. Neel to Earle Reynolds, 1 September 1953, 6/R, JVN.

25. Duncan McDonald to James V. Neel, 21 August 1953, 6/V, JVN.

26. Duncan McDonald to James V. Neel, personal letter, 21 August 1953, 6/V, JVN.

persisted, even if the comparison was between heavily and lightly exposed. "It may not now be significant," he said, referring to the stillbirth effect, "because of the tremendous loss of mothers involved, but it is there to almost the same extent just the same." McDonald's proposal to eliminate the controls would mean "the sacrifice of approximately 80 percent of our data. Since we knew in advance that we were looking for small differences, such a sacrifice of data might quite well reduce below the level of significance anything real which was there."[27]

McDonald's suggestion that the controls be thrown out, Neel said, also "surprised" him, since he considered his and Schull's approach "more conservative" than the earlier McDonald-Morton approach, which combined cities and incorporated all the material. "Now, suddenly, you spring to the other extreme and propose to completely discard all the 1s [unexposed] from consideration." The "sum and substance" of the matter was that "the perfect control population has not yet been made, but by discarding a sufficient amount of material one can reduce the significance of any finding." In any case, he said the differences in interpretation were relatively minor. "It is not as if we were claiming that the stillbirth effect or the birth weight effect were the results of irradiation." But he was "unwilling to discard a great mass of valid data just to eliminate these relatively minor differences."[28]

In any event Neel was now proposing to turn the dispute over to the Committee on Atomic Casualties. He, Schull, and Morton would present the entire debate to the committee on 13 September. "Your point of view should be ably represented by Morton. If the majority see it your way, then I've had it, and will make such changes as seem indicated. But if they do not, then I shall go ahead and send into *Science* substantially the manuscript which is now in your hands."[29]

Meanwhile Neel appealed to Grant Taylor under separate cover, saying that he was surprised at the "tone and content" of McDonald's criticism of the draft but otherwise agreed with the legitimacy of objections raised by staff members at Hiroshima.[30] He told Taylor that he and Schull were willing to consider minor revisions, but if McDonald had "very major objections" to the current draft of the paper, "then I have no alternative but to say to the Committee on Atomic Casualties: 'This is the way we think it should be presented; this

27. James V. Neel to Duncan McDonald, 31 August 1953, 6/R, JVN.

28. Ibid.

29. Ibid.

30. James V. Neel to Grant Taylor, 12 August 1953, 6/R, JVN.

is the viewpoint of our current geneticist in Japan; What is your advice in the matter?' "[31] Such a presentation proved to be necessary.

The day preceeding the 13 September 1953 meeting of the Committee on Atomic Casualties, Neel, Schull, and Morton had a session which began at 5 P.M. and "extended, one way or another, far into the night."[32] Summarizing the committee's response in a letter to McDonald five days later, Neel said that Morton was there "from beginning to end and very ably presented the viewpoints which you and he had developed. . . . We are in essential agreement that the differences in the way of looking at the data which exist at present between yourself and myself are probably no more than might be expected in the midstream of a complicated analysis." The differences had been given "a prominence which they do not at all deserve" by the fact that McDonald had decided to present the genetics data to the Japanese Genetics Society. "I certainly am in sympathy with the argument for the presentation to them; it's the matter of timing. At any rate we all feel strongly that now that the commitment has been made it should certainly be honored."[33]

With this in mind, then, he suggested that McDonald revise his planned presentation to conform to the Neel-Schull analysis, for both Morton and the Committee on Atomic Casualties had, in the end, endorsed the Neel-Schull approach. "It was the clear consensus of the group . . . that your approach tended, as you are well aware, to minimize the picture, this perhaps somewhat more than was justifiable at present." Some concessions would be made to McDonald, Neel noted. A sentence pointing out the urban versus rural background difference between the exposed and control groups would be inserted, though Neel personally was "very doubtful" of the significance of this difference. Some mention would also be made of the fact that comparing groups 1 and 2 to groups 4 and 5 produced different results than comparing groups 2 and 3 to groups 4 and 5. Both concessions related to McDonald's fundamental objection to the use of the control population. Neel said that McDonald's draft presentation to the Japanese group would be expeditiously revised by Morton to conform to the consensus reached by the Committee on Atomic Casualties. The committee had reaffirmed the importance of simultaneous and basically *identical* publication in both Japan and the United States. McDonald therefore was compelled to follow the Neel-Schull approach in his public talk.[34]

31. Ibid.

32. James V. Neel to Duncan McDonald, draft letter, 18 September 1953, 6/R, JVN.

33. Committee on Atomic Casualties, "Minutes," 13 September 1953, NAS. Also, James V. Neel to Duncan McDonald, draft letter, 18 September 1953, 6/R, JVN.

34. Ibid.

The product of all these negotiations was a brief report in the 6 November 1953 issue of *Science*, "The Effect of Exposure to the Atomic Bombs on Pregnancy Termination in Hiroshima and Nagasaki: Preliminary Report." Its authors, listed on the publication, were Neel, Schull, McDonald, and Morton, as well as other ABCC staff members, including cytogeneticist Masuo Kodani, pediatricians John Wood and Ray Anderson, and statistician Richard Brewer. The Japanese NIH was not mentioned or thanked in the acknowledgments—a matter of some embarrassment after the essay appeared.

In the *Science* piece, data were presented only for mothers under age thirty-five and for nonconsanguineous marriages. The comparisons were between offspring both of whose parents fell in the two highest exposure groups (4 and 5) and offspring both of whose parents fell in the lightly exposed or unexposed groups (1 and 2). Group 3 parents, for whom radiation exposure levels were most problematic, were not included in this preliminary analysis. The unexposed—the original controls for the study, who constituted approximately two-thirds of the total sample—were lumped with the lightly exposed, those present in Hiroshima or Nagasaki at the time of the bombings but exposed at a distance greater than 2,544 meters.[35] The indicators chosen for analysis in this report were stillbirth, sex ratio, birth weight, and major malformation. Anthropometric measurements were not included nor was there any reference to spontaneous abortion (Neel, Schull, McDonald, et al. 1953b, 538).

The authors reported that sex-ratio effects were significant in Nagasaki, even when the comparison was between heavily exposed and lightly exposed (rather than heavily exposed and not exposed). Sex-ratio differences were not present, however, in the more numerous Hiroshima data. There was no evidence of effects on malformation, they said, but there was some suggestive evidence of effects on stillbirth, in male infants born in Hiroshima to exposed mothers. But the problem of maternal effects, the authors noted, cast doubts on the genetic importance of this.

Two passages in the text appeared in italics in this initial publication. The first dealt with the possible confounding factors in the etiology of the selected indicators. *"Each of these possible indicators of genetic damage is also influenced by a number of other factors; there are no unique yardsticks of a genetic effect."* This same sentence had been underlined in the original draft of the paper in July. The second sentence had also appeared in the earlier draft, but it

---

35. Many of the control parents had a "rural background," a complicating factor that Neel and Schull tried to assess as a possible variable in their analysis. A greater proportion of the control parents, called "group 1," were repatriates from Korea and Manchuria who had settled in Hiroshima or Nagasaki after 1945.

had not been underlined. In the published version it too was partially italicized: "Late somatic effects of irradiation [cataracts and leukemia, at the time] have been established as occurring in a small fraction of heavily irradiated persons." Other "more subtle" effects might also exist, and since maternal health affected pregnancy outcome, *"extreme caution must be exercised in the genetic interpretation of any apparent effect of irradiation mediated solely by the mother"* (Neel, Schull, McDonald, et al. 1953b, 538). The stillbirth effects in Hiroshima, therefore, might be somatic rather than genetic effects.

The paper mentioned the peculiar birth weight results: "Induced mutations might be expected to impair the metabolic processes of the fetus and so decrease birth weight, i.e., the children of the more heavily irradiated parents might be expected to weigh less at birth." Instead, at least in Nagasaki, maternal and paternal exposure both appeared to increase birth weight, at the level of statistical significance. The explanation for this was "not readily apparent." One possibility was parity—that is, the number of pregnancies a woman has experienced. Birth weight increases with the number of pregnancies, so the age of the survivors, who tended to be older than the controls, might be responsible for the difference. But this analysis had truncated the data to eliminate mothers more than thirty-five years old, which should have ameliorated the influence of parity to some degree.

The possibility of somatic selection, raised in the July draft and articulated again by Neel in his correspondence with McDonald in September, was not mentioned in the public text. But Neel continued to wonder if those who survived tended to be larger and taller (assuming a relationship of body mass to sensitivity to radiation) than the average Japanese, with the result that survivors had a tendency to larger offspring. He would "bet my money that this [birth weight] is a parity effect." Yet he had been "wrong just often enough" to wonder if something else might be involved. He reminded McDonald after the Committee on Atomic Casualties meeting in the fall of 1953 that he had earlier suggested the possibility of somatic selection in the survivors. "You rejected this possibility as having absolutely no basis in known fact. May I refer you to the current issue of *Science*. You will find an article reporting a reasonably high correlation between body weight and the time to onset of the symptoms of radiation sickness following therapeutic irradiation. You will also find in that article a statement that the later the onset of radiation illness, the less severe the illness. I would like to repeat the suggestion that we ask the medical program to check its data to see whether the survivors are any taller or heavier than the average Japanese" (Court-Brown and Mahler 1953).[36]

36. Neel's discussion of this result, James V. Neel to Duncan McDonald, draft letter, 18 September 1953, 6/R, JVN.

Neel, Schull, and their coauthors concluded the essay in *Science* by stating that the findings were "consistent with what is known of the radiation genetics of a wide variety of plant and animal material, including *Drosophila* and mice." This was the sentence that disturbed McDonald and Morton, but it made it into the article. "There is no indication from this study of any 'unusual' sensitivity of human genes to irradiation," the authors said. The sentence had been through several contortions in the course of the various drafts of the paper, and the awkward syntax reflected the awkwardness of the problem.

The ABCC geneticists felt compelled to assess the relationship of their findings to the results with experimental organisms. Yet the results could not reasonably be portrayed as revealing that human sensitivity to radiation was the same as mouse or fly sensitivity. As McDonald pointed out, the radiation "effects" detected at Hiroshima and Nagasaki might even be compatible with the view that radiation had no effect at all on human genes. Morton, in reviewing the summer draft in 1953, particularly attacked the suggestion that the results indicated that the relative mutability of human and *Drosophila* or mouse genes might be equivalent. The data did not "justify any conclusion about the relative mutability of mice, *Drosophila* and men," Morton objected.[37] Yet publication without some treatment of the problem of comparison with the mouse and fly data would have been unthinkable: The study was a consequence of the results with experimental organisms and it was expected to address the cross-species comparison problem. Finding the proper constellation of words—to say neither more nor less than warranted—was difficult.

In the end, the favored solution was the use of double negatives. McDonald's paper in Japan concluded that there was "no indication from this study that the sensitivity of human genes to radiation is not comparable to that of experimental plants and animals used in radiation experiments."[38] The *Science* article said, somewhat more boldly, that the findings were "consistent with what is known of the radiation genetics of a wide variety of plant and animal material" and they gave "no indication of any 'unusual' sensitivity of human genes to radiation" (Neel, Schull, McDonald, et al. 1953b). Similarly, a National Academy of Sciences committee, assessing the preliminary results, concluded that there was "no evidence that the sensitivity of man to the genetic

37. Newton Morton to Record, 28 July 1953, drawer 3, ABCC Research Committee 1952–1953, NAS.

38. James V. Neel et al., "The Effect of Exposure of Parents to the Atomic Bombs on the First Generation Offspring in Hiroshima and Nagasaki: Preliminary Report," 27 October 1953, 6/V, JVN, 11.

effects of irradiation is not of the same order of magnitude as established in other, experimentally accessible organisms."[39]

The 1953 *Science* piece made its strongest claim in relation to sex-ratio effects. It said that sex ratio was "the [indicator] most biologists would probably feel has the largest genetic component in its etiology" (Neel, Schull, McDonald, et al. 1953b). As a consequence, the 1953 paper is generally remembered for having shown significant sex-ratio effects, effects that were later discounted when the full analysis was presented in 1956 (and revived in 1958, and rescinded in 1966; Schull and Neel 1958; Schull, Neel, and Hashizume 1966).

The National Academy of Sciences provided official summaries of these results to the press at the same time that the *Science* piece was made available. News organizations were asked to present these summaries in their entirety. A press packet sent out 7 November 1953 contained four statements, differing in length—four-page, two-page, one-page, and one-paragraph—distributed with the request that the news organization "incorporate" whichever statement was most appropriate to its needs—in other words, use this exact text, these exact words.[40]

As a consequence of its brevity, this preliminary publication glossed over or ignored many difficulties of the study and failed to include information about possible alternative analyses. The detailed exploration of the results in all their ramifications would have to wait another three years, until Neel and Schull could complete their statistical analyses and work through their conclusions.

THE MONOGRAPH

In August 1956 Neel and Schull formally presented their final results at the First International Congress of Human Genetics at Copenhagen. Later that same year, they published these results, slightly amended, in a monograph, once tentatively named "The Children of Hiroshima," but published as *The Effect of Exposure to the Atomic Bombs on Pregnancy Termination in Hiroshima and Nagasaki*.[41]

Neel said in late 1952 that while they called their preliminary analysis a

39. G. W. Beadle, D. R. Charles, C. C. Craig, L. H. Snyder, and Curt Stern, "Statement of the Conference on Genetics to the Committee on Atomic Casualties," 11 July 1953, 6/R, JVN.

40. "Release to Press," 7 November 1953, 6/V, JVN.

41. In 1991, Neel and Schull published a collection of papers dealing with the ABCC genetics study, under the title *The Children of Atomic Bomb Survivors*.

"pilot study," he suspected that much of what they did that fall would be "the definitive treatment."[42] He was correct in that many of the decisions made in the preliminary study were more or less incorporated into the final study.

By 1956, Neel and Schull accepted McDonald's contention that the unexposed parents were not comparable to the exposed group because of the "rural background" of many of the controls. The control population included many repatriates from Manchuria, Korea, and Formosa who had moved into Hiroshima and Nagasaki after the war. It was unclear what impact this would have on the indicators under study, but in the final publication Neel and Schull proposed that "any difference between category 1 [unexposed] and categories 2 through 5 [exposed at various levels] must be viewed with reservations" (Neel and Schull 1956a, 71). Thus they dismissed their control population and essentially stated that differences between it and the exposed populations could be attributed to causes other than radiation exposure.

In deciding to deemphasize their control data, Neel and Schull accepted the idea that the comparisons between lightly exposed and heavily exposed could be expected to reveal genetic effects of radiation more clearly than comparisons between exposed and unexposed, because this comparison minimized extraneous variables. The unexposed parents were not an "adequate control," and the "only meaningful comparison" involved infants both of whose parents were present in the cities at the time of the bombings (Neel and Schull 1956a, 87). If the radiation effects were real, they would be as real (if less dramatic) when the lightly exposed were compared to the heavily exposed, they reasoned.

As Neel had pointed out to McDonald in 1953, the total genetic effects of the bombs were expected to be small. Would such small effects persist at the statistically significant level even when both groups being compared had been exposed to radiation from the bombings? The idea of a threshold dose, below which no radiation effects could be expected, had some currency in the 1950s, but most geneticists rejected the possibility. Presumably, if there were no threshold dose, genetic effects would occur on a continuum. Heavy exposure might produce more mutations than light exposure, but light exposure would produce some mutations. The presence of mutations in the lightly exposed would make detecting a significant genetic effect even more difficult.

More importantly, the decision to emphasize comparisons of the lightly and heavily exposed placed a heavy burden on the dosimetry estimates. Miscalculating a survivor's exposure level either way could diminish the validity of the comparison. Classifying a heavily exposed survivor as lightly exposed

42. James V. Neel to Duncan McDonald, 11 September 1952, 6/Q, JVN.

would dilute the comparison, as would categorizing a lightly exposed survivor as heavily exposed. And dosimetry was the most complicated analytical problem they faced.

When the final results were published in 1956, the "significant" or near-significant sex-ratio effect detailed in the 1953 paper had disappeared. Further analysis of the entire body of data did not substantiate the earlier results, and Neel and Schull were left with a 241-page monograph that showed, essentially, very little. They began by describing the administrative framework of the study. They explained the pregnancy registration program, the home visits, the nine-months examinations, and the autopsy program. They laid out the story of the genetics program in the first two chapters and posed their basic question: "Can there be observed . . . any differences between the children born to parents, one or both of whom were exposed [to the bombings] and the children born to suitable control parents?" They then suggested some of the difficulties encountered in attempting to answer this question (Neel and Schull 1956a, section 6.1).

There followed a historical chapter on the "peopling of Japan," focusing on Nagasaki's special status as Japan's "gateway to the West." They postulated that the large numbers of foreigners who lived in Nagasaki over the centuries might have affected the genetic characteristics of the city's population, which might mean that the results in Hiroshima and Nagasaki would not be comparable. There were of course other reasons the two cities might not compare, including the differences in the radiation mix of the two bombs, the different topography of the two cities, and differences in climate (specifically humidity) which might affect radiation exposure levels.

Neel and Schull then began an exploration, by indicator, of the results of their work. They accorded a chapter to each of the six indicators selected for analysis—sex ratio, malformation, stillbirth, birth weight, growth at nine months, and death in the nine-month period following birth. They explained or at least mentioned many of the alternative statistical manipulations they had considered in the course of their analyses, and they compared the data on each indicator in several ways.

Neel and Schull's monograph joined a small group of other studies focused on the problems of the genetic effects of radiation in man. Work with human populations was (and is) almost exclusively limited to those exposed to radiation as a result of the atomic bombs, in medical therapy, or as an occupational hazard. Some researchers had also attempted to track the genetic impact of living in an area with a high level of natural background radiation, such as the monazite sands area in Kerala, India, or some sections of upstate New York. Background radiation, however, might have teratongenic effects on the developing embryo, so these studies were not directly comparable to the

ABCC study.[43] Neel and Schull did feel compelled in their 1956 text to explicitly compare their analysis to two surveys of radiologists carried out by Stanley Macht and Philip Lawrence (1955) and James Crow (1955; Turpin, Lejeune, and Rethore 1956, 201).

Stanley Macht was director of the department of radiology at Washington County Hospital in Hagerstown, Maryland. Philip Lawrence was chief of the familial studies unit of the Division of Public Health Methods in the U.S. Public Health Service, also in Hagerstown. In October 1951, Macht and Lawrence, with support from a research grant from the National Institutes of Health, launched a survey of radiologists and other physicians to detect possible genetic effects of radiation. They mailed almost eight thousand questionnaires, about half to radiologists and half to physicians in medical specialties unlikely to involve exposure to radiation. Their questionnaire asked the physicians to indicate how many years they had been regularly exposed to radiation through X-ray diagnosis, fluoroscopy, radium therapy, or use of radioisotopes, and whether they had ever been exposed to levels greater than the accepted tolerance levels. They also asked respondents to describe their reproductive history—number of children, miscarriages, congenital defects, and stillbirths. The physicians were also asked if there were any anomalies present in their immediate families or those of their spouses. The few female radiologists who responded—62, compared to 2,717 men—were excluded from the analysis because maternal and paternal irradiation pose different biological problems (Macht and Lawrence 1955).[44] Two months earlier James Crow, a geneticist at the University of Wisconsin, had mailed similar surveys to 1,027 radiologists and 1,036 controls drawn from pathologists listed in the Directory of Medical Specialists. Crow learned about the Macht and Lawrence survey after his own survey had been distributed. Since the Macht and Lawrence survey was larger, Crow delayed publication of his results to coincide with theirs. The two studies differed in one important respect. Macht and Lawrence explained the purpose of their survey in a cover letter. Crow merely asked for the participants' cooperation without explaining his purposes (1955).

The Macht and Lawrence results indicated that the offspring of exposed

43. The highest levels of natural radiation were believed to be in Travancore, India, on ground containing monazite sands. The gonad dose in this area was estimated at between 10 and 20 rads per individual in a thirty-year period (Gentry, Parkhurst, and Bulin 1959; Wesley 1960).

44. One of the physicians surveyed strongly objected to the plan, and wrote back to urge Macht to "get another horse to ride." "I might wish that your enthusiasm and your ambitions might arouse your interest in avenues of radiology that would be more stimulating than a record of mishaps." E. H. Skinner to Stanley Macht, 27 November 1951, sent by Stanley Macht to Vilma Hunt, 1985, to whom I am indebted for this reference.

fathers had higher rates of abnormalities, though not at the statistically significant level. They divided their data in various ways—by birth order, parity, age
of mother—and found trends consistent with genetic expectations. Their sex-
ratio results, given that they used only results from irradiated fathers, were
peculiar. They found fewer male births for irradiated fathers, while in theory
male irradiation might results in fewer female offspring due to the induction
of dominant lethal mutations on the paternal X chromosome. They concluded
that the differences detected between the offspring of radiologists and those of
other physicians were not "of large magnitude" and should not be "viewed
with alarm." Yet these abnormalities were visible in "the first generation of
offspring, and visible first generation effects represent only a small fraction of
the total damage that may have been inflicted" (Macht and Lawrence 1955,
466).

Crow's results, published simultaneously in the same journal, indicated
no significant differences in rates of fetal and infant death in the offspring of
radiologists and pathologists, though again, any differences were in the expected direction. Crow did not seek information about abnormalities. He concluded by pointing out twice, in very similar language, that the negative or
inconclusive findings of his study had "little bearing on the question of the
over-all genetic damage to the human population in future generations," particularly since "only a very small fraction of genetic damage due to radiation
appears in the first generation" (Crow 1955, 470).

Neel and Schull, unimpressed by Macht and Lawrence's results, felt they
had to address the study, "inasmuch as their study will be widely quoted" (Neel
and Schull 1956a, 201). They dismissed the Macht and Lawrence findings on
two grounds. First, they questioned whether radiologists with an "unusual"
reproductive history might be more highly motivated to reply to the questionnaire—though Macht and Lawrence had explicitly addressed this issue. Questionnaires were returned immediately by 74.1 percent of the radiologists but
only 53.8 percent of the controls. In this first round rates of fetal death and
congenital defects were 3.10 percent higher in the offspring of radiologists
than in those of non-radiologists. But in the second round of responses, many
more controls reported abnormalities, and this difference dropped to 0.76 percent. This was "a striking and disturbing discrepancy," Neel and Schull noted.
It might mean that radiologists with abnormal offspring were more likely to
answer the questionnaire immediately than were control parents with abnormal
offspring. This could be a significant source of bias (Neel and Schull 1956a,
200–204).

Second, they questioned the prevalence, according to the Macht and Lawrence sample, of two of the reported abnormalities—a blood disease and a lung

disease—among the offspring of radiologists.[45] These conditions, not "usually regarded as malformations," accounted for one-third of the observed difference in the two sets of offspring. Neel and Schull noted that Macht and Lawrence had included as abnormalities such conditions as Tay-Sachs disease, which would not have been detectable in the Hiroshima-Nagasaki data. Tay-Sachs is an inherited degenerative disease that appears before age five. It would not have been detectable in the ABCC data, since infants were examined only at birth and at nine months. Macht and Lawrence had also included minor malformations and scored multiple defects in an individual separately. While Macht and Lawrence felt there were some trends indicating differences between the two groups, Neel and Schull concluded, "we do not feel they have established this fact" (1956a, 200–204).

In the final two chapters of the 1956 book, "Recapitulation" and "Permissible Inferences," Neel and Schull concluded that their study could not provide strict quantitative estimates of radiation risks. The study would justify itself, they said, if it proved to be of some value in assessing radiation risks in human populations. They closed by noting that nuclear energy was a necessary new technology, given the energy needs in human society. Public concern about radiation exposure "*must* find a frame of reference" (emphasis in original), they said. "As coal and oil resources dwindle, increasing recource will be had to atomic fuel as a source of energy" (1956a, 217).

Radiation exposure from fallout, or from some future nuclear war, would almost certainly increase the rate of mutation in human populations, Neel and Schull acknowledged in their final chapter, and there was a "high probability (but not certainty) that under the conditions of Western culture such mutations [would] act to the detriment of the populations concerned." Much of the additional "genetic load" to be expected in human populations might come from radiation, but much might also come from "one of a number of dysgenic influences at work in human populations." War and "differential birth rates" might also act to the detriment of human populations, they said, and "the quantification of these influences is just as unsatisfactory as that of radiation" (1956a, 216). "It will seem to many that to attempt to phrase the problem in those terms is hopelessly impractical. At the moment, yes." But Neel and Schull believed "very substantial advances" in human genetics would soon be possible. They expected "great progress" on the problem of the genetic effects of radiation on human populations, and "on the time scale of human evolution, within a relatively short period of time" (1956a, 217).

45. These were erythroblastosis fetalis and atelectasis of the lungs.

They agreed that radiation exposure should be minimized for all persons "until we have a clearer idea of just how harmful these effects are." But without this better understanding, it would be "as unfortunate, on the one hand, to deny the possibility that low doses are dysgenic at all as it is, on the other hand, to assert that a serious threat to the genetic integrity of mankind is involved." They added that "even if the present limits are ultimately found to be too high, there are few who would argue that in the period it takes to establish that fact with certainty, man will have suffered serious genetic harm" (1956a, 216–17).

RESPONSES TO THE MONOGRAPH

It was difficult not to be impressed with the complexity of the ABCC genetics study, and reviewers generally recognized that Neel, Schull, and their coauthors faced profound problems in both data collection and data analysis. No one hailed the book as a major breakthrough, though one reviewer in the *Eugenics Quarterly* judged it a "fitting monument to those killed or maimed in the first dawn of the atomic age" (Allen 1957, 105). James Crow's review in the *American Journal of Human Genetics* noted that "a study of this magnitude and with as many unavoidable uncertainties must depend on a number of arbitrary choices." He considered the assignment of survivors to the five groups arbitrary and said that "Class 5, curiously, includes all who had any symptoms . . . irrespective of distance" from the hypocenter. The "extreme conservatism" of the analysis resulted in "some loss of data," he said. Crow also questioned the decision to approach the study as a test of the hypothesis that no genetic effects had occurred at Hiroshima and Nagasaki—the null hypothesis for this case "In my opinion a change of emphasis from one of testing the null hypothesis to one of estimating as well as possible the range of possible effects would have provided more useful information from the data." But, he added, the burden of such a criticism fell on the investigator who proposed a different approach: "For the authors state that any investigator who wishes to try other analytical methods may obtain a duplicate set of the IBM cards on which the original analysis was based" (Crow 1957b, 225).

Crow said the authors were "unnecessarily critical" of Muller, and he questioned their extreme reluctance to provide quantitative estimates of radiation effects in human populations. Neel and Schull feared that such figures would be immediately "used" and that they could be misunderstood by persons "not thoroughly indoctrinated in genetics and unfamiliar with the shaky basis of the primary assumptions." But Crow argued that "when there are decisions to be made provisional and inaccurate quantitative estimates are better than none at all" (Crow 1957b, 226). Another reviewer, in the 1 November 1957 *Science,* echoed Crow's criticisms, saying that "educated estimates, even though they may be quite inaccurate, are better than no estimates, and this is especially true at present when standards of permissible dosages of radiation are being set" (Steinberg 1957, 932).

It is hard to know what to make of this promotion of the values of inaccurate estimates by the book's reviewers, though I suspect it was a frustrated response to the general style of Neel and Schull's text and its circumspection and defensiveness. Neel and Schull were conservative in their wording and approach throughout the text. Their cautious dismissal of data and comparisons that seemed to suggest genetic effects was all legitimate and proper, but also vaguely troubling given the real public concern about the biological effects of radiation. The text reads as a defensive argument, the authors poised and ready for attack. In 1961, such an attack finally appeared.

This attack took the form of an alternative analysis of the ABCC data, carried out by Paul de Bellefeuille, a Canadian pediatrician who reanalyzed the ABCC data and found significant genetic effects for some indicators in some groups. He published two papers in *Acta Radiologica* suggesting that Neel and Schull had attempted to conceal genetic effects clearly indicated by their data. De Bellefeuille recognized that the dosimetry classifications were a problem, though his understanding of the nature of this problem was incomplete. He pointed out that the sex-ratio effect, which seemed significant in the 1953 paper, had disappeared in the 1956 monograph, and he was extremely suspicious of the ABCC "shielding correction" that reclassified some survivors into lower exposure categories. "One may thus well ask if the shielding correction constitutes an improvement for any reason other than that it affords a more reassuring view of the long-term effects of the atomic bombing." De Bellefeuille also asserted incorrectly that Neel and Schull "measure[d] the statistical weight of the exposure of one parent without taking into account whether the other parent was exposed or not." He therefore decided to analyze the offspring of parental pairs in which one parent was heavily or lightly exposed and the other not exposed at all. This analysis, he reported, revealed significant effects for sex ratio, stillbirth, neonatal death, and total loss (a compilation of all effects including sex ratio; de Bellefeuille 1961a, 75–78). De Bellefeuille concluded that his "new analysis of the data based on the separate study of the exposure of each parent alone" brought out "definite indications of genetic ill-effects of atomic radiation, at a high level of statistical significance" (1961, 155).

Neel considered de Bellefeuille's reanalysis "really quite vicious." He and Schull responded with an attack on de Bellefeuille's statistical expertise and reasoning, in a published paper that Neel said contained "probably the strong-est language I have ever used in print, but the situation seemed to call for it."[46] Indeed, this rejoinder was lively. Neel and Schull said that they had made a "particular effort" in their 1956 book to "present the data in such a fashion that others so minded could undertake alternative forms of analysis." But "never . . . did it occur to us that the data could be so misused as in the recent commen-tary of de Bellefeuille." They assessed de Bellefeuille's comments on shielding as an "obvious slur on the personal integrity of the present authors" and said these comments were typical of the entire piece. "While we would not contend that the precise scheme we have employed . . . is beyond reproach, we feel it represents a step in a necessary direction. We can assure Dr. de Bellefeuille that, the labor involved being what it is, we did not run two analyses, one with and the other without allowance for shielding, and then select the one which most reassured us." De Bellefeuille "demonstrates a flair for novelty both in arithmetic and statistical technique," engages in a "crudely contrived effort to undermine the reader's credulity [sic] in the data," and makes statements that are "patently false," Neel and Schull added. They then answered his assess-ment point by point, in relation to sex ratio, malformation, stillbirth, neonatal death, and the autopsy data, showing that they had either already carried out the analyses he suggested or that such analyses were not legitimate (Neel and Schull 1962).

Another criticism came somewhat belatedly from the great but eccentric J. B. S. Haldane, who suggested in 1964 that the Neel-Schull findings could not be trusted because the Japanese could not be trusted. Haldane was skeptical of the entire enterprise, but he was particularly suspicious of the Japanese, and—an amazing hypothesis—suggested that Japanese participants mali-ciously distorted the data in an effort to trick the United States into using atomic weapons again. He said clerks employed by the ABCC might have changed the coding on the IBM cards so that malformed offspring born to heavily exposed survivors disappeared from the data. Their goal, he said, was to make it appear that no genetic effects had occurred and thus to encourage the use of the atomic bomb on populations in the Soviet Union or the United States (Haldane 1964).[47] His imagined scenario revealed his ignorance of ABCC operations, of the social and political context of radiation effects in postwar Japan, and of the nature of Japanese involvement in the study.

46. James V. Neel to Keith Cannan, 7 June 1962, drawer 19, Genetics—1959–1968 #1, NAS. See also de Bellefeuille (1961) and Neel and Schull (1962).

47. Haldane expressed hostility toward the Japanese, noting that the "politest are, in my opin-ion, the most suspect" and suggesting that "about a million Japanese would die very happily in exploding an atomic bomb in an American city, if the opportunity offered" (Haldane 1964).

CONCLUSION

At the time that the 1956 Neel and Schull monograph was in final draft, a United Nations scientific committee was preparing a report on the genetic effects of radiation. The UN Scientific Committee on the Effects of Atomic Radiations (UNSCEAR), which first met in March 1956, found the report on the ABCC genetics project extremely useful. It could be cited as evidence that genetic effects from low levels of radiation were unlikely and could be used to refute the horror stories that occasionally appeared in the popular press or in advocacy literature. The UNSCEAR report included a brief discussion (one paragraph) of the Neel-Schull results, noting that though these results were negative, they did provide some evidence of a "lower limit for the representative doubling dose for human genes, at least for the dominant mutations which would have been observed by these authors" (United Nations 1958, 184).

The 1956 book closed one chapter in the ABCC's efforts to track the genetic effects of radiation, but it did not end the search.

CHAPTER TWELVE

The ABCC and the RERF

A t the laboratories of the Radiation Effects Research Foundation in Hiroshima, in the Hijiyama Hill buildings still known by local cab drivers as "ABCC," a large freezer holds blood samples taken from survivors and their children and grandchildren. The DNA in these samples may eventually help answer the question that first drew Neel to Japan: Did radiation from the atomic bombs cause an increase in heritable mutations in the offspring of survivors? Even the most spectacular techniques of contemporary molecular biology have not yet been able to resolve that question.[1]

The 1956 results did not convince Neel, Schull, or the scientific commu-

---

1. Many of the relevant papers, as well as the full manuscript of the original 1956 publication, are contained in Neel and Schull, eds., *Children of the Atomic Bomb Survivors: A Genetic Study* (1991). See particularly Neel and Susan Lewis, "The Comparative Radiation Genetics of Humans

nity at large that the question had been answered. Because there was an early sex-ratio effect, which persisted, though below the level of statistical significance, in the later data, the ABCC continued to collect information about sex ratio in the offspring of exposed and control populations after the collection of data on other traits ceased in early 1954. By 1958, the number of children born to parents either or both of whom had been exposed to radiation was falling. Only about 15 percent of the births in the two cities involved such parents. The survivors as a group were getting older, moving out of the child-bearing years, and were also moving out of Hiroshima and Nagasaki.[2]

Partly in response to these demographic changes, the ABCC adopted a new approach. Three cohorts of children were selected, one consisting of children born to heavily exposed survivors, the other two of controls matched by age, sex, and city. The parents of one group of controls had not been exposed to the bombings at all; the parents of the other control group had been distally or lightly exposed, at distances of at least twenty-five hundred meters from hypocenter. The children were then followed through life, so that genetic effects not obvious at birth or in childhood could be detected. The groups were compared for mortality rates and other possible signs of radiation effects.[3]

In 1968, with the encouragement of Neel and Schull, the ABCC began a cytogenetic study of the frequency of balanced structural rearrangements of chromosomes in the children of survivors (Awa et al. 1968). In 1972 the ABCC initiated an electrophoretic study of proteins, in an effort to detect recessive mutations in the children of survivors (Neel, Satoh et al. 1980). In 1990 Neel, Schull, and other authors reported on the rates of malignant tumors in the children of survivors (Yoshimoto et al. 1990). All of these efforts to detect mutations in the children of bomb survivors have been inconclusive. Some have hinted at radiation effects—particularly after various reinterpretations of the statistical data. But as Neel observed in 1989, "with no statistically significant findings, we might only be manipulating the noise in the system" (Neel, Schull et al. 1989, 47).

Meanwhile ABCC work on the somatic effects of radiation exposure in the survivors themselves began to take more of the ABCC's resources. Significant effects began to appear by 1949, with the detection of radiation cataracts,

---

and Mice," *Annual Review of Genetics* 24 (1990):327–62, reprinted in Neel and Schull (1991, 451–86).

2. The critical sex-ratio papers are Schull and Neel, "Radiation and the Sex Ratio in Man" (1958), and Schull, Neel, and Hashizume, "Some Further Observations on the Sex Ratio among Infants Born to Survivors of the Atomic Bombings of Hiroshima and Nagasaki" (1966).

3. See the summary of this work in Neel and Schull, eds., *The Children of the Atomic Bomb* (1991, 4–5).

and by 1952 higher rates of leukemia had been amply documented. Other malignancies, including breast, lung, stomach, and thyroid cancer, were established as effects of radiation exposure over the next twenty-five years. More recently, the RERF reported cardiovascular effects of radiation on survivors. Small head size and mental retardation were documented in children exposed in utero, and delays in growth and development were demonstrated in persons exposed as children. The ABCC-RERF has been able to document a high relative risk for leukemia in persons with exposure levels of 200 rads or more. These findings of late somatic effects have been one important basis for worker protection legislation around the world.[4]

As the scientific work has changed, so has the administrative organization of the ABCC. The most important change came in 1975, when the organization was renamed the Radiation Effects Research Foundation. With the new name and administrative organization, funding for the research on the survivors was to be provided equally by the United States and Japan. My attention in this book has focused on the Occupation and immediate post-Occupation period, because this was the first phase of the genetics project. I want here briefly to tell the story of the ABCC since 1956, suggesting how the organization has adapted to changing political climates and to changing scientific technologies, which have shifted attention from the diseased body to molecular structure.

THE FRANCIS REPORT

In late 1955 Keith Cannan, chairman of the NRC Division of Medical Sciences, appointed a committee, chaired by University of Michigan epidemiologist Thomas Francis, to assess the scientific work of the ABCC. Francis was joined by Seymour Jablon, an NRC staff member in the statistics division, and Felix E. Moore, head of the biostatistics program at the NIH National Heart Institute. The review was prompted partly by an unpublished report on death

4. There are two established contemporary radiation protection standards, one issued by the International Commission on Radiation Protection in 1977, and the other by the National Council on Radiation Protection and Measurements in 1987. Alice Stewart has criticized the 1977 ICRP standards for their dependence on RERF results, suggesting that the Hiroshima data has played an unjustifiably large role in determining worker protection standards (Stewart 1971, 361f; Stewart 1985, 55; and Stewart and Kneale 1988). The 1977 standards are still widely followed, as Jacob Shapiro points out in his *Radiation Protection: A Guide for Scientists and Physicians*, noting that the new U.S. Nuclear Regulatory Commission standards in 1992 were to be based on the ICRP's 1977 system of radiation dose limitations. The NCRP considered setting lower limits for radiation exposure in their 1987 report but generally stayed with the ICRP recommendations (Shapiro 1990, 341–64).

rates in Hiroshima, prepared by two ABCC staff members there. Flaws in the epidemiological reasoning that shaped the report's conclusions disturbed readers at the NAS and NRC and suggested that the ABCC staff needed more guidance.[5]

The members of the Francis Committee found, in their visit to Hiroshima, that the organization and its work were in chaos. Morale was low and the scientific work was proceeding at an unsatisfactorily slow pace. "Whatever the blueprint or plan of operation originally provided, it seems either to have been indistinct, or has been lost in the successive changes of staff," committee members told Cannan in their report. The professional staff was not committed to the project, and the departments were not communicating or working together. At the same time, the organization was devoting significant time and resources to the training of Japanese physicians, a worthy goal but a "diversion from the objective" of the ABCC.[6]

Despite their reservations about the existing operation, however, the three members of the Francis Committee strongly favored continuing the effort in Hiroshima and Nagasaki. The data already collected were "uniquely valuable" and might prove to contain important information about radiation effects. They recommended that a Unified Study Program be instituted, pulling together all the efforts of the ABCC. Most importantly, they suggested that the ABCC identify a clearly defined population for long-term study, a group that would not change even if new survivors came forward to express their willingness to participate. The survivor population was constantly shifting in both Hiroshima and Nagasaki as new cases of exposure were discovered, and as former participants migrated out of the two cities. This created a great deal of confusion. The Francis Committee recommended that the study population should be drawn from those who had identified themselves as survivors in the 1950 Japanese census. While the master file would continue to track persons not included in that census, the data on them should be handled separately.[7]

The committee favored the creation of an epidemiological detection network in which "monitors," selected from among the subjects of the study, would report weekly on the health status of fifteen or twenty other subjects. The monitor would be responsible for persons in his immediate neighborhood, with whom he had personal relationships. Such a network could provide imme-

---

5. The suggestion that the report was suppressed came in a conversation with Gilbert Beebe, March 1988, Hiroshima. See also Beebe (1988).

6. Francis Report, 6 November 1955, NAS.

7. Ibid.

diate information about both minor and significant illness, and about migration of survivors and controls.[8]

The Francis Report outlined specific changes needed in each of the ABCC research programs and suggested that relations between the study subjects and the ABCC needed to be improved. Many subjects were dropping out of the program, partly because the examination itself was a costly inconvenience. It was "strongly recommended" that participants be paid the day's wages lost for their examination at the ABCC. "The excessive character of loss is seen in the ME-55 Program, where less than 50 percent of the original population was brought in for a third examination." Physicians who spent time trying to convince families to permit the ABCC to conduct autopsies should also be compensated, the committee members said.[9] They concluded their report by saying that it was "apparent" that the ABCC program was "bogged down because of deficiencies in understanding of purpose, in ideas, in initiative and in outlook. There has been a lack of investigative leadership and stimulating support."[10]

The Francis Report was controversial—it angered many staff members in Hiroshima—and not all its recommendations were followed. The proposal for monitors, for example, was rejected in favor of home visits or health surveys of industrial sites where many survivors were employed. Later both the home visits and the site visits were abandoned because of their cost, and history taking continued to be a matter of interviews at the twice-yearly examinations (Beebe 1979, 186). But the proposals in the Francis Report did lead to productive changes in the organization of the program in Japan. The life span study of a hundred thousand persons, the mortality study of twenty-eight hundred persons exposed in utero and controls, and the mortality study of the offspring of survivors were all consequences of the suggestions of the Francis Committee.

Perhaps of equal importance, the committee's observations on staffing problems led to the establishment of a collaborative relationship with the Yale School of Medicine and the University of California at Los Angeles. Under this agreement, faculty members from the two schools could spend a year or more working at the ABCC in Japan without giving up their positions and leaving the academic loop. This made it possible for American scientists to contribute time and energy to the ABCC without sacrificing academic careers

8. Ibid.
9. Ibid.
10. Ibid.

in the United States.[11] The impact of this arrangement on the ABCC was substantial. Many American scientific faculty members spent time with the ABCC after 1957, and the practice continues in the present at the RERF. Yale particularly continues to have a special relationship to the survivors and the medical studies that have been conducted on them—a relationship that began in the weeks after the bombings when the first Joint Commission teams arrived in Japan.

Five of the seven medical officers assigned to Japan for the Joint Commission were either Yale faculty or Yale medical alumni. One of the first pathologists assigned to the ABCC in 1948, Dr. William J. Wedemeyer, was a member of the Yale faculty. Francis himself was a Yale Medical School graduate. Other Yale faculty who worked with the ABCC in Japan included Stuart Finch, Lawrence R. Freedman, Kenneth G. Johnson, and Benedict R. Harris, and Robert Jay Lifton was a Yale professor of psychiatry when he published his psychological study of atomic bomb survivors, *Death in Life: Survivors of Hiroshima* (Lifton 1968; "Yale and the ABCC" 1968).

One of the most important of those to come to the ABCC from Yale was George Darling, who served as ABCC director longer than any other person and whose leadership made the recommendations of the Francis Report a reality. Darling, a professor of human ecology, came to the ABCC on leave of absence in 1957, and stayed as director for fifteen years, until December 1972. When NAS president Detlev Bronk asked him to accept a temporary appointment at ABCC in 1957, Bronk said the ABCC was "important for the future of mankind." But one of Darling's American friends said the ABCC was "so scientifically and diplomatically difficult" that it was "a wild goose chase."[12] Darling went to Japan anyway and made the best of it.

Perhaps more than his predecessors, Darling emphasized the ABCC's status as a form of "international science." Until his tenure, virtually all ABCC publications appeared only in English. Darling instituted bilingual publication of ABCC annual and technical reports and the use of bilingual forms throughout the ABCC. He oversaw the initiation of the ABCC Adult Health Study, the creation of a Department of Medical Sociology, and the establishment of the cytogenetics laboratory. In 1967 Darling engineered the return of biologi-

---

11. The agreement with Yale called for the university to staff the position of chief of medicine at the ABCC. It was an informal agreement between Cannan and Yale chairman of internal medicine Paul Beeson. The Yale professor would then bring over with him teams of younger physicians, primarily from Yale, for one- or two-year terms ("Yale and the ABCC" 1968).

12. Darling recounted this in the Atomic Bomb Casualty Commission Annual Report, 1 July 1971–30 June 1972, iv.

cal materials and autopsy records to the Japanese, thus helping to resolve longstanding tensions over the American "confiscation" of survivors' body parts.

As discussed in my chapter 2, Americans working with the survivors in the early years after the bombings—in both the Joint Commission and the ABCC—collected slides, photographs, autopsy records, and organ specimens that had been compiled by Japanese scientists and physicians. Much of this material was sent to the Armed Forces Institute of Pathology in Washington, D.C., and not returned to the scientists and physicians who had originally obtained it. The "confiscation" of Japanese biological data and specimens persisted as a source of tension in Japanese-American relations through the 1960s, playing a role in narratives that described the Americans as ghoulishly hoarding survivor body parts. Darling personally arranged to have all materials collected by the ABCC returned to Japan, to a new research institute in Hiroshima. Partly as a result of his efforts, the materials collected by the first Joint Commission were returned to Japan six years later, in 1973.[13]

Meanwhile the ABCC's financial status began to be an operating problem. Darling pointed out in 1967 that the financial organization was anachronistic: Why should the United States be providing most of the funding more than twenty years after the end of the war? (Beebe 1979, 202). He appealed to the Japanese government for more funding for the ABCC after a 1971 revaluation of the yen had devastated the ABCC budget. Both the Atomic Energy Commission and the National Academy of Sciences authorized additional funding in response to this revaluation, but the Japanese government did not. "I wish I could report that the Government of Japan had taken advantage of this opportunity to offer to assume the cost of additional segments of the operating program by increasing its appropriation," Darling wrote. He could not report this, however, and only because of quick action by the AEC and the NAS had the ABCC "avoided a crisis."[14]

Talks on reorganization and a new financial structure for the ABCC formally began in 1974, though Darling had by this time been campaigning for such talks for at least seven years. The "fiscal crisis" provoked by the revaluation of the yen in 1971 "carried the threat of closure" and brought the United

---

13. On the return of Japanese materials from the Armed Forces Institute of Pathology, see "JANC Listing of Materials for Transfer to Japanese Government" and other related materials, including press clippings, box 12, A-Bomb manuscripts, OHA.

14. George Darling, Atomic Bomb Casualty Commission Annual Report, 1 July 1971–30 June 1972, x.

States and Japan to the bargaining table (Beebe 1979, 202). Yet as Beatty has pointed out, the Japanese government was not particularly interested. The issue was only partly financial. Japanese leaders were also reluctant to take responsibility for the ABCC, since, as one American participant in the negotiations observed, this would "also expose the government to its share of the annual round of criticism" of the ABCC (usually coinciding with the anniversary of the bombings) (cited in Beatty 1993, 227).

Despite such reservations, the government of Japan and the Japanese National Institute of Health did agree to take on a larger share of the burden—both financial and political—for the studies of the survivors. The Radiation Effects Research Foundation was established on 1 April 1975 as a nonprofit foundation under Japanese Civil Law, pursuant to an agreement between the United States Department of Energy—successor to the AEC—and the Japanese National Institute of Health. Funding was to be provided equally by the two governments, instead of disproportionately by the United States, and scientific and administrative control were effectively balanced, if anything, in favor of the Japanese. Technically the organization is managed by a binational board of directors, with the assistance of a binational scientific council. In effect, however, since 1975, Japanese scientists and administrators have played a much larger role. Many American scientists, including Neel and Schull, remain actively involved with the RERF, but it can no longer be characterized as an American organization. It maintains a public image as a special form of "international science," and "evidence of the joint nature of the project" is "everywhere on exhibit: from jointly authored papers to carefully designed letterhead." (Beatty 1993, 231).

Many of those who criticized the ABCC in the early years and throughout its operations attributed the organization's problems in Japan to the lack of Japanese control. In the decade before the reorganization in 1975, the ABCC found that Japanese medical investigators were increasingly unwilling to share data. This might suggest that Japanese control of, or at least more active involvement in, the ABCC would help resolve such problems. But as Gilbert Beebe has observed, the change in 1975 had no such effect. "All the old difficulties face RERF as they did ABCC." Beebe concludes that earlier control by the Japanese would have made no difference, but I think he overlooks the importance of history to the way people respond to an institution (Beebe 1979, 202).

While ABCC officials promoted the organization as "collaborative" or cooperative as early as 1947, it was always managed at the top by Americans. Participants in the study knew that this was the case and did not mistake the involvement of Japanese pediatricians or midwives as a sign that the ABCC was a Japanese project. It is also important to recognize that, in this case, the "nationality" of the project—its status as "American"—was extremely important, much more so than it would have been had the Americans been studying a more neutral medical phenomenon. By 1975, when the organizational structure did genuinely change, the RERF retained the cultural meaning of the ABCC, and it would be expecting too much to witness a complete change in public reaction in only four years. The RERF sponsored a massive fortieth anniversary celebration in March 1988 in Hiroshima and Nagasaki—a celebration I attended—thus emphasizing the continuity between the RERF and the ABCC. For Hiroshima cab drivers, at least, the RERF does not seem to exist: thirteen years after the name change, I found that telling a cab driver to take me to the "ABCC" produced immediate understanding, while "RERF" produced merely a puzzled look.

In 1986 the need to estimate mutation frequency in atomic bomb survivors and their offspring became the justification for Department of Energy involvement in the effort to map the entire human genome, the Human Genome Project. In November 1991 the RERF sponsored a "germline mutagenesis workshop" at which international experts and RERF staff scientists agreed that the survivors' children "may be the most important source of information related to human mutation studies." The offspring of survivors, workshop attendees agreed, provided "unique scientific opportunities." RERF chief of research James Trosko predicted that "emerging molecular techniques for detecting mutations" place the genetics project on the threshold of solving the genetic questions at last. Demonstrating a statistically significant difference in mutation rates between the children of the exposed and control populations would still be an "enormous task," comparable to a "second human genome project," Trosko noted. Yet new technologies on the horizon might make such a study feasible, and the RERF is currently investigating the possibility ("Mutagenesis Experts Brainstorm at RERF" 1991).

CONCLUSION

While the ABCC's primary articulated purpose was to study the medical effects of irradiation on the survivors, the organization was also expected to have an "impact on the future course of Japanese medicine and science." This was the "responsibility in history" of which "the staff of the commission and members of the parent committee [were] acutely conscious," according to Philip Owen of the National Research Council.[15] Owen felt that the ABCC was a

15. Indeed, Owen suggested that this provided a "compelling motivation [for participation]" especially among the senior scientific members of the group." He pointed out that a State Department publication referred to Japan as one of the "strategic channels for the spread of United States

"pilot plant" that, if "brought to its full potential," would "write in a significant paragraph at least in the history of international science of two great nations."[16]

I think Owen was correct in interpreting the ABCC as an important study in international science, not for revealing the independence of science from culture but rather for demonstrating how nationality and culture (in both the political and intellectual senses) frame scientific decisions. The story of this organization illustrates how science works when collaboration involves groups with different conceptions of the problems to be resolved. Scientists from different nations commonly perceive scientific problems differently, particularly when those problems involve national issues of public health, autonomy, or military security.

The nuclear accident at Chernobyl, for example, like the bombings at Hiroshima and Nagasaki, created a natural experiment with human populations. And in the debate over radiation effects at Chernobyl, nationalist tensions emerged. Those attempting to assess the impact of radiation on these populations faced political controversy, government secrecy, and problems with estimating dosimetry—both because the exact amount of radiation released at Chernobyl is still in question and because exact records of people's movements and activities after the accident are not available. As at Hiroshima and Nagasaki, the question of the long-term medical effects of the radiation at Chernobyl became the subject of scientific disagreements that split along nationalist (or regionalist) lines. The site of the accident is in the Ukraine, near the border with Byelorussia. Ukrainian and Byelorussian scientists, concerned about the welfare of people living in the contaminated area, challenged scientists in Moscow on the clean-up plan, on radiation exposure standards, and on the proper zones of evacuation. In a familiar story, the scientists in the republics claimed that the Soviet government scientists were underestimating long-term health risks and concealing information about risks and about the death toll (Clines 1990).

The Japanese often expressed similar concerns. The estimation of health risks when one nation is "responsible" and the other victimized may commonly break down along these lines. The Japanese perception that the ABCC (or, rather, the United States government) was not entirely forthcoming in releasing information about the biological effects of radiation was frequently accurate. The information was a state secret because it was tied to a weapon

---

scientific information and ways of thinking in the Far East" (International Science Policy Survey Group 1950).

16. See Philip Owen to Robert Keith Cannan, "Memorandum," 7 July 1950, Unitarian Medical Education Mission to Japan—1951, NAS.

that shaped U.S. Cold War strategies. But Japanese responses were also consistent with a narrative of exploitation and secrecy that characterizes situations involving indefinite risks and international or interethnic cooperation. Radiation—invisible, mysterious, with medical effects that can take decades to appear—is a particularly potent focus of such tensions.

CHAPTER THIRTEEN

Conclusions

T he survivors of the atomic bombings at Hiroshima and Nagasaki are largely silent in the archival memos, correspondence, and committee reports that have formed the primary documentary basis of this book. They appear as coded numbers, as cases, as categories ("heavily exposed," "lightly exposed"), but they do not speak. They are anonymous points of data in the scientific publications and internal correspondence.

My study of the ABCC has focused on the literature of anonymous survivors—on the scientific reports and the internal memos in which their personal identity was excised, consciously, even necessarily, so that a different meaning could be derived from their experiences. The ABCC made the suffering of the survivors scientific, a fact, a truth about the world, in ways that a dead baby could not. The ABCC translated the suffering into a different language, a pow-

erful language that could be interpreted and used in the distant world of workers' legal claims, legislative agendas, and energy policy programs.

This translated text did not capture the complete experience of the bombings. It was limited and like most translations, distorted. But it did create us-
able public knowledge. *What happened* to the survivors—the slow and invisible internal pathologies of their bodies over the decades—was gradually made
visible and real by the science of the ABCC.

Anthropologist Hugh Gusterson's study of weapons scientists at Lawrence
Livermore Laboratory explores how physicists describe bodies destroyed by
nuclear weapons as though they were machines. The body is presented "as a
set of components which undergo mechanical interactions with blast waves
and glass fragments" (Gusterson 1992, 216–17). For the physicians and biologists studying the effects of radiation on the survivors, the bodies of the survivors were not mechanical objects interacting with glass or stone, but clusters
of organs (gonads, blood, heart) interacting with a form of energy. Each reaction in each organ was a separate point on a statistical graph, and the meaning
of any individual event was unclear. Only in the aggregate could the data reveal
biological (and political) truth.

The difficulty of this task was demonstrated in the ABCC genetics project,
which was an effort to make the invisible mathematically visible. Had there
been mutations? If they could not be read conclusively in the dead or malformed bodies of newborns, where could they be read? In the statistical data
that would reveal that they existed? In the molecular record of survivors'
DNA? The biological effects of the bombings—both genetic and somatic—
had no specific signature. The survivors and their children suffered and died
in ways no different from other human beings. The reality of their illness, then,
was statistical, and in the absence of statistical significance radiation effects,
literally, did not exist. The biological effects of radiation on the survivors were
made real not by their existence in the survivors' bodies, but by the 967 scientific papers published by ABCC scientists from 1947 until 1974, papers detailing the effects of radiation on life span, aging, fertility and sterility, immuno-
logical response, cancer, cardiovascular disease, and other pathologies.[1] In this
same process, much of the survivors' experience disappeared.

This irony—of science as both the creator and destroyer of truth—has
been the subject of this book. The bombings and their aftermath had many
meanings; the ABCC focused on one of those meanings—the bomb as source
of radiation—but other meanings intruded, interfered, and had to be suppressed, controlled, sometimes denied. This process of suppression and control

1. See *Bibliography of Published Papers of the Atomic Bomb Casualty Commission 1947–
1974* (Hiroshima and Nagasaki: Atomic Bomb Casualty Commission, 1974).

was manifest in the day-to-day practices of ABCC employees, in ABCC policies, and in the minutes of the advisory meetings and private correspondence
that have formed the documentary record of my text.

I would not suggest that the scientists involved should have devoted themselves to empathizing with the survivors, or that they were individually indifferent to the personal pain endured by many survivors. Americans involved
with the ABCC were often affected personally by the suffering they witnessed.
But that suffering was explicitly excluded from the scientific study. The process of selection, distancing, and separation that characterized so much of the
ABCC's work may be the central business of science. Japanese critics of the
ABCC interpreted this distancing as inhumane: how could the suffering of the
survivors be irrelevant to the work of the Americans? From the perspective of
American participants, however, responding emotionally to the suffering of the
survivors was acceptable social practice but not "good science."

As my account shows, however, the emotional meaning of the survivors
remained, stubbornly, refusing to disappear, lingering in the margins of AEC
directives and calculations of exposure levels. The screen through which the
survivors' experiences were filtered was imperfect. The symbolic importance
of the survivors was present in the scientific analysis of their bodies, the policies that guided the collection of data, and the interactions of scientists and
subjects. The data that drew on their suffering was a manifestation of the cultural meaning of the bomb and the special historical place of those whose
damaged bodies could reveal its implications for humanity.

I would add that the phenomenon of distancing takes many forms. Some
survivors distanced themselves emotionally from what they witnessed in the
minutes, hours, and days after the bombings. The tragic consequences of the
bombings became invisible even to the sufferers, who often experienced a
rapid psychic closing off. As the survivor quoted earlier stated, human beings
in horrible pain became "mere substance." Making the suffering at Hiroshima
and Nagasaki disappear, then, was both an immediate act of psychological
work, and a long-term act of scientific work.

Slowly, piece by piece, we still struggle to make sense of the events in
these two cities in August 1945. They do not have a single meaning, and to see
any one of them, it seems, one must try to limit the glare from the others:
the bomb as source of biological change, as military necessity, as scientific
achievement for the community of physicists, as diplomatic tool, or as inhumane Cold War politics.

In the fall of 1993, the *New York Times* reported that both the Soviet Union
and the United States engaged in secret radiation experiments with human subjects in the 1940s and 1950s. The Soviets exploded an atomic bomb near forty-
five thousand troops and thousands of civilians as part of a military exercise.

Their goal was to understand how troops would perform under conditions of nuclear war (Simons 1993). Meanwhile the United States was conducting similar tests, in which large numbers of American troops were exposed to radiation, and under the auspices of the Atomic Energy Commission, conducting human experimentation to determine the biological effects of radiation. Eighteen American civilians were injected with plutonium in the 1940s, to help determine safe worker exposure levels. As many as six hundred other persons were subjected to other experiments with radiation in the same period (Healy 1993).

These newspaper stories, bits of data in the wake of the unraveling of the Cold War, starkly reveal the military importance of information about the biological effects of radiation. Bodies scarred by radiation were a critical military commodity, so important that their acquisition—through diplomacy and politics (as in the ABCC), atmospheric testing, or human experimentation—justified drastic measures. The Cold War was a war, like any other war, and radiation was as potent an enemy as communism. The atomic bomb created bodies damaged by the invisible force that haunted the Cold War and its warriors, in ways that we are only now beginning to be able to see.

It has not escaped my notice that historical analysis is another forum in which the problem of the meaning of the bombs and of the survivors' experiences can be worked out. I now find that I have joined that chorus of persons who have tried to place the experiences of the survivors at Hiroshima and Nagasaki in a meaningful context. My attention here focuses on the survivors in the Cold War, and the scientific interpretation that translated their experiences into data for worker protection legislation and other uses. But those two days in August 1945 cast a shadow on this text just as they cast a shadow on the work of the Atomic Bomb Casualty Commission. The bombings of Hiroshima and Nagasaki changed what it means to be human, for the bomb opened the door to annihilation. The bomb joins that other central horror of the 1939–1945 war—the organized mass killing in Nazi Germany—to weigh heavily on all residents of the twentieth century. As Robert Jay Lifton has eloquently suggested, we are all survivors, in one way or another, of these events.[2]

2. "Identification guilt, like the bomb's lethal substance itself, radiates outward. In Hiroshima this 'radiation' moved from the dead to the survivors to ordinary Japanese to the rest of the world" Lifton (1968, 498–99).

Bibliography

ARCHIVAL SOURCES

The most important and extensive archival records of the Atomic Bomb Casualty Commission are held by the National Academy of Sciences in Washington, D.C., in thirty-three file drawers. The NAS has prepared a guide to the collection which lists topics and files. But materials relating to the ABCC are also in other NAS files, including the records of the Division of Medical Sciences and the director's office files.

Of crucial importance to the study of the genetics project are the extensive and well-organized personal papers of James V. Neel. Currently these are in Ann Arbor. Neel has a commitment, however, to donate them to the American Philosophical Society in Philadelphia. Some of these papers have also been promised to the Houston Academy of Medicine at the Texas Medical Center, but at the time I examined them it was unclear which papers would be sent to Philadelphia and which to Houston.

Also temporarily in Ann Arbor are several boxes of papers saved by William J. Schull. These will be sent to the Houston Academy of Medicine where some of Schull's other papers are located. The Houston Academy of Medicine also holds the papers of Grant Taylor, William Moloney, and Carl Tessmer. The extensive photo collection in Houston is particularly valuable, though, unfortunately, many of the photos are undated and lack identification.

Records of the Advisory Committee on Biology and Medicine of the Atomic Energy Commission are crucial to the ABCC story. These are in the Department of Energy archives in Germantown, Maryland. Many relevant committee meeting minutes, however, are still classified.

The Bentley Historical Collection of the University of Michigan holds relevant papers of Thomas Francis, and the Rockefeller Archive Center in Pocantico Hills, New York, holds the papers of John C. Bugher.

The National Archives in Washington has some materials relevant to the State Department's interest in the ABCC. The National Library of Medicine has some Joint Commission papers and a valuable oral history interview with Shields Warren (valuable partly because Warren's papers do not survive). In addition, the Armed Forces Institute of Pathology Archives in Washington has complete detailed records of the transfer of biological materials back to Japan in 1967 and 1973, as well as some materials from the Joint Commission and the ABCC. The Wilmington College Peace Resource Center,

Wilmington, Ohio, holds useful Japanese materials and translations relating to the bombings and the Japanese response.

Abbreviations

APS — American Philosophical Society, Philadelphia, Pennsylvania.

BHL — Michigan Historical Collection, Bentley Historical Library, Ann Arbor, Michigan.

DOE — Papers of the Advisory Committee on Biology and Medicine, RG 326, U.S. Atomic Energy Commission, Department of Energy Archives, Germantown, Maryland.

GT — Papers of Grant Taylor, Archives of the Houston Academy of Medicine, Texas Medical Center, Houston, Texas.

HAM — Archives of the Houston Academy of Medicine, Texas Medical Center, Houston, Texas.

JCB — Papers of John C. Bugher, Rockefeller Archives Center, Pocantico Hills, New York.

JVN — Personal papers of James V. Neel, Ann Arbor, Michigan.

NAS — ABCC Collection, Archives of the National Academy of Sciences, Washington, D.C.

NLM — National Library of Medicine, Bethesda, Maryland.

OHA — Otis Historical Archives, National Museum of Health and Medicine, Armed Forces Institute of Pathology, Washington, D.C.

WJS, AA — Personal papers of William J. Schull, Ann Arbor, Michigan. These papers will eventually be transferred to HAM.

WJS, HAM — Personal papers of William J. Schull, ABCC Collection, Archives of the Houston Academy of Medicine, Texas Medical Center, Houston, Texas.

Ann Arbor, Michigan. Bentley Historical Library. Papers of Thomas Francis.
Ann Arbor, Michigan. Bentley Historical Library. Papers of James V. Neel.
Ann Arbor, Michigan. Bentley Historical Library. Papers of William J. Schull.
Bethesda, Maryland. National Library of Medicine.
Bethesda, Maryland. National Library of Medicine. Shields Warren oral history.
Germantown, Maryland. Department of Energy Archives. Papers of the U.S. Atomic Energy Commission, Advisory Committee on Biology and Medicine, [RG 326, boxes 3217–18].
Houston, Texas. Archives of the Houston Academy of Medicine, Texas Medical Center. Papers of William J. Schull.
Houston, Texas. Archives of the Houston Academy of Medicine, Texas Medical Center. Personal papers of Grant Taylor, ABCC Collection.
Houston, Texas. Archives of the Houston Academy of Medicine, Texas Medical Center. MS. collection 73, diary of William Moloney.
Philadelphia, Pennsylvania. American Philosophical Society. Papers of Curt Stern.

Philadelphia, Pennsylvania. American Philosophical Society. MS. Collection 49, American Society of Human Genetics.

Pocantico Hills, New York. Rockefeller Archive Center. Papers of John C. Bugher.

Washington, D.C. Archives of the National Academy of Sciences. Organized Records. See "Survey of ABCC and RERF Papers in the NAS Archives," National Academy of Sciences, 19 November 1981.

Washington, D.C. Archives of the National Academy of Sciences. Unorganized Records. ARM N-14-4 to N-20-5. See "Survey of ABCC and RERF Papers in the NAS Archives," National Academy of Sciences, 19 November 1981.

Washington, D.C. Otis Historical Archives, National Museum of Health and Medicine, Armed Forces Institute of Pathology. Papers of the Joint Army-Navy Commission.

GENERAL REFERENCES

A-Bomb Victims Association. 1953. *Life after the atomic bombing: Notes of atomic bomb victims.* Hiroshima. Pamphlet.

Agawa Hiroyuki. 1957. *Devil's heritage.* Translated by John M. Maki. Tokyo: Hokuseido Press.

Akizuki Tatsuichirō. 1981. *Nagasaki 1945: The first full-length eyewitness account of the atomic bomb attack on Nagasaki.* Translated by Keiichi Noyata; edited by Gordon Honeycombe. London: Quartet Books.

Allen, Gordon. 1957. Review. *Eugenics Quarterly* 4, no. 2 (June):105.

Alperovitz, Gar. 1965. *Atomic diplomacy, Hiroshima and Potsdam: The use of the atomic bomb and the American confrontation with Soviet power.* New York: Simon and Schuster.

American Men and Women of Science: A biographical directory. 1955. 9th ed. Vol. 2, *The Biological Sciences.* New York: J. Cattell Press, R. R. Bowker.

―――. 1962, 10th ed. *The Biological Sciences.* New York: J. Cattell Press, R. R. Bowker.

Amrine, Michael. 1959. *The great decision: The secret history of the atomic bomb.* New York: Putnam.

Atomic Bomb Casualty Commission. 1974. *Bibliography of published papers of the Atomic Bomb Casualty Commission, 1947–1974.* Hiroshima and Nagasaki: Atomic Bomb Casualty Commission.

Atomic Bomb Victims' Written Notes Editorial Committee. 1953. *Genbaku ni ikite: Genbaku higaisha no shuki* [Life after the atomic bombing: Notes of atomic bomb victims]. Hiroshima Pamphlet.

Auxier, John A. 1977. *Ichiban: Radiation dosimetry for the survivors of the bombings of Hiroshima and Nagasaki.* Oak Ridge, Tenn.: Energy Research and Development Administration, Technical Information Center.

Awa, A. A., A. D. Bloom, M. C. Yoshida, Neriishi S., and P. Archer. 1968. A cytogenetic survey of the offspring of atomic bomb survivors. *Nature* 218:367–68.

Barker, Rodney. 1985. *The Hiroshima maidens: A story of courage, compassion and survival.* New York: Viking.

Beatty, John. 1987. Weighing the risks: Stalemate in the Classical/Balance Controversy. *Journal of the History of Biology* 20, no. 3: 289–319.

———. 1991. Genetics in the atomic age. In *The expansion of American biology,* edited by Keith Benson, Jane Maienschein, and Ronald Rainger, 284–324. New Brunswick, N.J.: Rutgers University Press.

———. 1993. Scientific collaboration, internationalism, and diplomacy: The case of the Atomic Bomb Casualty Commission. *Journal of the History of Biology* 26, no. 2:205–31.

Beck, H. L., and G. DePlanque. 1968. The radiation field in air due to distributed gamma-ray sources in the ground. U.S. Atomic Energy Commission Report.

Beebe, Gilbert W. 1979. Reflections on the work of the Atomic Bomb Casualty Commission in Japan. *Epidemiological Reviews* 1:184–210.

———. 1988. Thoughts on the anniversary of ABCC/RERF. *RERF Newsletter* 14 (1 March):16–17.

Beebe, Gilbert W., H. Kato, and C. E. Land. 1978a. The hypothesis of radiation-accelerated aging in the mortality of Japanese A-bomb survivors. In *Late biological effects of ionizing radiation: Proceedings of the symposium on the late biological effects of ionizing radiation held by the International Atomic Energy Agency in Vienna, 13–17 March 1978,* 1:3–27. Vienna: International Atomic Energy Agency.

———. 1978b. Studies of the mortality of A-bomb survivors: Mortality and radiation dose, 1950–1974. *Radiation Research* 75:138–201.

Beer, Lawrence W. 1984. *Freedom of expression in Japan: A study in comparative law, politics, and society.* Tokyo: Kodansha International.

Benedict, Ruth. 1946. *The chrysanthemum and the sword: Patterns of Japanese culture.* Boston: Houghton Mifflin and Co.

Blackett, P. M. S. 1949. *Fear, war and the bomb: Military and political consequences of atomic energy.* New York: Whittlesey House.

———. 1962. Atomic heretic. In *Studies of war: Nuclear and conventional,* 73–77. Westport, Conn.: Greenwood Press.

Blair, Henry A., ed. 1954. *Biological effects of external radiation.* New York: McGraw-Hill.

Blot, W. J., and H. Sawada. 1972. Fertility among female survivors of the atomic bombs at Hiroshima and Nagasaki. *American Journal of Human Genetics* 24:613–22.

Boyer, Paul. 1985. *By the bomb's early light: American thought and culture at the dawn of the atomic age.* New York: Pantheon.

Braibanti, Ralph. 1984. The MacArthur shogunate in Allied guise. In *Americans as proconsuls: United States military government in Germany and Japan, 1944–1952,* edited by Robert Wolfe, 77–91. Carbondale: Southern Illinois University Press.

Brandt, Allan M. 1985. Racism and research: The case of the Tuskegee syphilis study. In *Sickness and health in America: Readings in the history of medicine and public health,* 2d ed., rev., edited by Judith Walzer Leavitt and Ronald L. Numbers, 331–43. Madison: University of Wisconsin Press.

Braw, Monica. 1986. *The atomic bomb suppressed: American censorship in Japan 1945–1949.* Malmö, Sweden: Liber Førlag.

Brecher, Ruth, and Edward Brecher. 1969. *The rays: A history of radiology in the United States and Canada.* Baltimore, Md.: Williams and Wilkins.

Bugher, John C. 1952. Delayed radiation effects at Hiroshima and Nagasaki. *Nucleonics* 10, no. 9 (September): 18–21.

Burchett, Wilfred. 1983. *Shadows of Hiroshima.* London: Verso Editions.

Carlson, Elof Axel. 1981. *Genes, radiation and society: The life and work of H. J. Muller.* Ithaca, N.Y.: Cornell University Press.

Caufield, Catherine. 1989. *Multiple exposures: Chronicles of the radiation age.* Chicago: University of Chicago Press.

Charles, Donald R., Joseph A. Tihen, Eileen M. Otis, and Arnold B. Grobman. 1960. Genetic effects of chronic X-irradiation exposure in mice. U.S. Atomic Energy Commission Research and Development Report, UR-565. [Rochester, N.Y.]: University of Rochester, Atomic Energy Project.

Chu, E. H. Y., N. H. Giles, and K. Passano. 1961. Types and frequencies of human chromosome aberrations induced by X-rays. *Proceedings of the National Academy of Sciences* 47:830.

Chugoku Shimbun and the Hiroshima International Cultural Foundation. 1983. *The meaning of survival: Hiroshima's 36-year commitment to peace.* Hiroshima: Chugoku Shimbun.

CIOMS. Council for the International Organization of Medical Sciences. 1991. *International guidelines for ethical review of epidemiological studies. Midwifery* 7:42–44.

Clark, G. L. 1936. A 1936 survey of the biological effects of X-irradiation. *Radiology* 26:295–312.

Clines, Francis X. 1990. A new arena for Soviet nationalism: Chernobyl. *New York Times,* 30 December.

Cogan, D. G., S. F. Martin, and S. J. Kimura. 1949. Atomic bomb cataracts. *Science* 110:654–55.

Cogan, D. G., S. F. Martin, S. J. Kimura, and H. Ikui. 1950. Opthalmologic survey of atomic bomb survivors in Japan, 1949. *Transactions of the American Ophthalmological Society* 48:62–87.

Coleman, Samuel K. 1990. Riken from 1945 to 1948: The reorganization of Japan's Physical and Chemical Research Institute under the American Occupation. *Technology and Culture* 31:228–50.

Committee for Compilation. See Committee for the Compilation of Materials on Damage Caused by the Atomic Bombs in Hiroshima and Nagasaki.

Committee for the Compilation of Materials on Damage Caused by the Atomic Bombs in Hiroshima and Nagasaki. 1981. *Hiroshima and Nagasaki: The physical, medical and social effects of the atomic bombings.* Translated by Eisei Ishikawa and David L. Swain. New York: Basic Books.

———. 1985. *The impact of the A-bomb, Hiroshima and Nagasaki, 1945–1985.* Tokyo: Iwanami Shoten.

Compton, Arthur H. The atomic crusade and its social significance. 1947. *Annals of the American Academy of Political and Social Science,* January, 249.

Coughlin, William James. 1952. *Conquered press: The MacArthur era in Japanese journalism.* Palo Alto, Calif.: Pacific Books.

Court-Brown, W. M., and R. F. Mahler. 1953. The X-ray dose-body size relationship in radiation sickness. *Science* 118:271–72.

Cousins, Norman. 1949. Hiroshima four years later. *Saturday Review of Literature,* 17 September, 8.

Cremin, Lawrence A. 1964. *The transformation of the school: Progressivism in American education, 1876–1959.* New York: Vintage Books.

Crow, James F. 1955. A comparison of fetal and infant death rates in the progeny of radiologists and pathologists. *American Journal of Roentgenology* 73 (March):467–71.

———. 1956. The estimation of spontaneous and radiation-induced mutation rates in man. *Eugenics Quarterly* 3:201–8.

———. 1957a. Possible consequences of an increased mutation rate. *Eugenics Quarterly* 4:67–80.

———. 1957b. Review: The effect of exposure to the atomic bombs on pregnancy termination in Hiroshima and Nagasaki. *American Journal of Human Genetics* 9, no. 3:224–26.

Curtin, Philip D. 1990. The end of the white man's grave? Nineteenth century morality in West Africa. *Journal of Interdisciplinary History* 21:63–88.

de Bellefeuille, Paul. 1961. Genetic hazards of radiation to man. Parts 1, 2. *Acta Radiologica* 56:65–80, 145–59.

DeSchweinits, G. E., and B. F. Baer. 1932. A note on radiational cataract. *Archives of Ophthalmology* 7:150–51.

Divine, Robert A. 1978. *Blowing on the wind: The nuclear test ban debate 1954–1960.* New York: Oxford University Press.

Dobzhansky, Theodosius. 1937. *Genetics and the origin of species.* New York: Columbia University Press.

Dower, John W. 1986. *War without mercy: Race and power in the Pacific war.* New York: Pantheon.

Drinnon, Richard. 1987. *Keeper of concentration camps: Dillon S. Myer and American racism.* Berkeley and Los Angeles: University of California Press.

Farber, S., L. K. Diamond, and R. D. Mercer. 1948. Temporary remissions in acute leukemia in children prolonged by folic acid antagonist, 4-aminopteroylalutamic acid (aminopterin). *New England Journal of Medicine* 238:787–93.

Feis, Herbert. 1961. *Japan subdued: The atomic bomb and the end of the war in the Pacific.* Princeton, N.J.: Princeton University Press.

———. 1966. *The atomic bomb and the end of World War II.* Rev. ed. Princeton, N.J.: Princeton University Press.

Finch, Stuart C., and Iwao M. Morihama. 1980. *The delayed effects of radiation exposure among atomic bomb survivors, Hiroshima and Nagasaki, 1945–79* Radiation

Effects Research Foundation Technical Report TR 16–78. Hiroshima: Radiation Effects Research Foundation.

Fogg, L. C., and S. Warren. 1941. Some cytologic effects of therapeutic irradiation. *Cancer Research* 1:649–52.

Folley, J. H., W. Borge, and T. Yamawaki. 1952. Incidence of leukemia in survivors of the atomic bomb in Hiroshima and Nagasaki. *American Journal of Medicine* 13:311–21.

Freireich, Emil J., and Noreen A. Lemak. 1991. *Milestones in leukemia research and therapy.* Baltimore, Md.: Johns Hopkins University Press.

General Headquarters. 1948. See General Headquarters, Supreme Commander of the Allied Powers, Public Health and Welfare Section.

General Headquarters, Supreme Commander of the Allied Powers, Economic and Scientific Section. 1952. *Mission and accomplishments of the Occupation in the economic and scientific fields.* Tokyo: General Headquarters, Supreme Commander of the Allied Powers.

General Headquarters, Supreme Commander of the Allied Powers, Public Health and Welfare Section. 1948. *Public health and welfare in Japan.* 2 vols. Tokyo: General Headquarters, Supreme Commander of the Allied Powers.

Genetics Conference, Committee on Atomic Casualties, National Research Council. Genetic effects of the atomic bombs in Hiroshima and Nagasaki. 1947. *Science,* 10 October, 331–33.

Gentry, J. T., E. Parkhurst, and F. B. Bulin, Jr. 1959. An epidemiological study of congenital malformations in New York state. *American Journal of Public Health* 49:497.

Giovanitti, Len, and Fred Freed. 1965. *The decision to drop the bomb.* New York: Coward-McCann.

Glasstone, Samuel. 1950. *Effects of atomic weapons.* Los Alamos, N.M.: Los Alamos Scientific Laboratory.

Gluck, Carol. 1983. Entangling illusions: Japanese and Ameican views of the Occupation. In *New frontiers in American–East Asian relations: Essays presented to Dorothy Borg,* edited by Warren I. Cohen. Studies of the East Asian Institute. New York: Columbia University Press.

Gofman, John W. 1981. *Radiation and human health.* San Francisco: Sierra Club Books.

Goldschmidt, Richard. 1952. The theory of the gene. *Cold Spring Harbor Symposia on Quantitative Biology* 16:1–11. Biological Laboratory, Cold Spring Harbor, N.Y.

Goldstein, L., and D. P. Murphy. 1929. Etiology of the ill-health in children born after maternal pelvic irradiation. *American Journal of Roentgenology* 22:322–31.

Goodman, David G., ed. 1985. *After apocalypse: Four Japanese plays of Hiroshima and Nagasaki.* Translated by David G. Goodman. New York: Columbia University Press.

Goodman, Grant K., ed. 1968. *The American occupation of Japan: A retrospective view.* Lawrence, Kans.: Center for East Asian Studies, University of Kansas.

Green, Earl L. 1968. Genetic effects of radiation in mammals. *Annual Review of Genetics* 2:87–120.

Greenberg, Daniel S. 1968. *The politics of pure science.* New York: New American Library.

Grobman, Arnold Brams. 1951. *Our atomic heritage.* Gainesville, Fla.: University of Florida Press.

Grosch, Daniel S., and Larry E. Hopwood. 1979. *Biological effects of radiations.* 2d ed. New York: Academic Press.

Gusterson, Hugh. 1992. Testing times: A nuclear weapons laboratory at the end of the Cold War. Ph.D. diss., Stanford University.

Hachiya Michihiko. 1955. *Hiroshima diary: The journal of a Japanese physician, August 6–September 30, 1945.* Translated and edited by Warner Wells. Chapel Hill: University of North Carolina Press.

Hacker, Barton C. 1987. *The dragon's tail: Radiation safety in the Manhattan Project, 1942–1946.* Berkeley and Los Angeles: University of California Press.

Haddow, A., ed. 1952. *Biological hazards of atomic energy.* Papers read at a conference convened by the Institute of Biology and the Atomic Scientists Association, October 1950. Oxford: Clarendon Press.

Haldane, J. B. S. 1938. *Heredity and politics.* New York: W. W. Norton and Co.

———. 1964. The harmless atom bomb? Review of *Changing perspectives on the genetic effects of radiation* by James V. Neel. *Journal of Genetics* 59 (August):86–88.

Hall, Eric J. 1984. *Radiation and Life.* New York: Pergamon.

Hardesty, Martha Ellen. 1986. Language, culture and Romaji reform: A communications policy failure of the Allied occupation of Japan. Ph.D. diss., University of Minnesota.

Harris, Harry. 1959. *Human biochemical genetics.* Cambridge: Cambridge University Press.

Healy, Melissa. 1993. Secret for years, 204 nuclear tests disclosed by U.S. *Philadelphia Inquirer,* 7 December.

Henshaw, Paul S. 1944a. Experimental roentgen injury. *Journal of the National Cancer Institute* 4:503–12.

———. 1944b. Leukemia in physicians. *Journal of the National Cancer Institute* 4:339–46.

———. 1991. The ABCs of the early days. *RERF Update,* winter 1991–92, 12–13.

Hersey, John. 1946. *Hiroshima.* New York: A. A. Knopf.

Hertwig, P. 1939. Zwei subletale recessive mutationen in der nachkommenschaft von rontgenbestrahlten mausen. *Arbarzt* 6:41–43.

Hewlett, Richard G., and Jack M. Holl. 1989. *Atoms for peace and war, 1953–1961: Eisenhower and the Atomic Energy Commission.* Berkeley and Los Angeles: University of California Press.

Hook, Ernest B. 1986. James Watt Mavor (1883–1963): A forgotten discoverer of radiation effects on heredity. *Perspectives in Biology and Medicine* 29:278–91.

Hosokawa, Bill. 1950. Atomic test tube: What happens to humans exposed to A-bomb radiation? With the help of Hiroshima's survivors, we are trying to find out. *Empire Magazine,* published by the *Denver Post,* 10 December.

———. 1951. ABCC of radioactivity. *Scene: The pictorial magazine,* January, 16–18.

Hubbell, H. H., T. D. Jones, and J. S. Cheka. 1969. *The epicenters of the atomic bombs: Reevaluation of all available physical data with recommended values.* U.S. Atomic Energy Commission Report, ABCC-TR-3-69. Hiroshima: Atomic Bomb Casualty Commission.

Ibuse Masuji. 1969. *Black rain: A novel.* Translated by John Bester. Tokyo: Kodansha International.

Imabori Seiji. 1959. *The era of A- and H-bombs.* Kyoto: Sanichi-Shobo.

International Science Policy Survey Group. 1950. *Science and foreign relations: International flow of scientific and technological information.* Publication 3860, General Foreign Policy, series 30. Washington, D.C.: Department of State.

Ishimaru, T., T. Hoshino, and M. Ishimaru. 1971. Leukemia in atomic bomb survivors, Hiroshima, Nagasaki. *Radiation Research* 45:216–33.

Jablon, Seymour. 1973. The origin and findings of the Atomic Bomb Casualty Commission. *Nuclear Safety* 14 (November–December):6.

Janis, Irving L. 1949. *Psychological aspects of vulnerability to atomic bomb attacks.* RAND Corporation Research Memorandum, RM-94. Santa Monica, Calif.: RAND Corporation.

———. 1951. *Air war and emotional stress.* New York: McGraw-Hill.

Japanese Broadcasting Corporation (NHK). 1977. *Unforgettable fire: Pictures drawn by atomic bomb survivors.* Translated by the World Friendship Center in Hiroshima. New York: Pantheon.

Johnson, Sheila K. 1988. *The Japanese through American eyes.* Stanford, Calif.: Stanford University Press.

Jones, James H. 1981. *Bad blood: The Tuskegee syphilis experiment.* New York: Free Press.

Jones, T. D., J. A. Auxier, J. S. Cheka, and G. D. Kerr. 1965. In-vivo dose estimates for A-bomb survivors shielded by typical Japanese houses. *Health Physics* 28:367–81.

Jungk, Robert. 1961. *Children of the ashes: The story of a rebirth.* Translated by Constantine Fitzgibbon. 1959; New York: Harcourt, Brace & World.

Kahn, E. J. 1961. Letter from Nagasaki. *New Yorker,* 29 July, 52–53.

Kato, H. 1971. Mortality in children exposed to the A-bombs while in utero, 1945–1969. *American Journal of Epidemiology* 93:435–42.

Kato, H., W. J. Schull, and J. V. Neel. 1966. A cohort-type study of survival in the children of parents exposed to atomic bombings. *American Journal of Human Genetics* 18:339–73.

Kawai Kazuo. 1960. *Japan's American interlude.* Chicago: University of Chicago Press.

Kevles, Daniel J. 1986. *In the name of eugenics: Genetics and the uses of human heredity.* Bekeley and Los Angeles: University of California Press.

Koya, Y. 1953. The program for family planning in Japan. *Eugenical News* 38:1–3.

Kubo, Mitsue. 1990. *Hibaku: Recollections of A-bomb survivors.* Translated by Ryoji Inoue, revised by Louise Herder. Coquitlam, B.C.: M. Kubo.

Kuklick, Henrika. 1991. *The savage within: The social history of British anthropology, 1885–1945.* New York: Cambridge University Press.

Kumar, Deepak. 1990. The evolution of colonial science in India: Natural history and the East India Company. In *Imperialism and the natural world,* edited by John H. MacKensie. Manchester: Manchester University Press.

Lapp, Ralph Eugene. 1958. *The voyage of the Lucky Dragon.* New York: Harper and Brothers.

Laughlin, J. S., and I. Pullman. 1957. *The biological effects of atomic radiation: Gonadal dose from the medical use of X-rays, preliminary report.* Washington: National Academy of Sciences, National Research Council.

Lea, D. E. 1947. *Actions of radiations on living cells.* Cambridge: Cambridge University Press; New York: MacMillan.

Leinfelder, P. J., and H. D. Kerr. 1936. Roentgen-ray cataract. *American Journal of Opthalmology* 19:739–56.

Levine, Robert J. 1986. *Ethics and regulation of clinical research.* 2d. ed. Baltimore, Md.: Urban and Schwarzenberg.

Liebow, Averill A. 1965. Encounter with disaster: A medical diary of Hiroshima, 1945. *Yale Journal of Biology and Medicine* 38, no. 2:61–239.

Life. 1949. The A-bomb's children: Study of half a million Japanese reveals the first delayed effects of atomic radiation. 12 December.

Lifton, Robert Jay. 1968. *Death in life: Survivors of Hiroshima.* New York: Random House.

Lilienthal, David Eli. 1963. *Change, hope and the bomb.* Princeton, N.J.: Princeton University Press.

Lindee, M. Susan. 1990. Radiation, mutation and species survival: The genetics studies of the Atomic Bomb Casualty Commission in Hiroshima and Nagasaki, Japan. Ph.D. diss., Cornell University.

Longino, Helen. 1990. *Science as social knowledge: Values and objectivity in scientific inquiry.* Princeton, N.J.: Princeton University Press.

Lorenz, E., C. Congdon, and D. Uphoff. 1952. Modification of acute irradiation injury in mice and guinea pigs. *Radiology* 58:863–77.

Lutman, F. C., and J. V. Neel. 1945. Inherited cataract in the B. genealogy: Occurrence of diverse types of cataracts in the descendants of one person. *Archives of Ophthalmology* 33:341–57.

Lyon, Mary F. 1961. Gene action in the X-chromosome of the mouse. *Nature* 190:372–73.

Macht, Stanley, and Philip S. Lawrence. 1955. National survey of congenital malformations resulting from exposure to roentgen radiation. *American Journal of Roentgenology* 73 (March):442–66.

Macklin, Madge Thurlow. 1941. Heredity and the physician. *Scientific Monthly* 52:56–67.

———. 1948. Making the doctor cancer-conscious. *Journal of the American Medical Association* 3(April):147–49.

Marshall, Eliot. 1992. Study casts doubt on Hiroshima data. *Science* 258 (16 October):394.

Mather, Kenneth. 1952. The long-term genetical hazard of atomic energy. In *Biological hazards of atomic energy,* edited by A. Haddow, 57–66. Oxford: Clarendon Press.

Matsumoto, Scott. 1954. Patient rapport in Hiroshima. *American Journal of Nursing* 54, no. 1 (January):69–72.

Mavor, James Watt. 1921. On the elimination of X-chromosome from the egg of *Drosophila melanogaster* by X rays. *Science* 54:277.

———. 1924. The production of non-disjunction by X rays. *Journal of Experimental Zoology* 39 (April):2.

———. 1952. *General biology.* 4th ed. New York: Macmillan Co.

McKusick, Victor A. 1975. The growth and development of human genetics as a clinical discipline. *American Journal of Human Genetics* 27:261–73.

Medical Research Council. See Medical Research Council (Great Britain), Committee on the Hazards to Man of Nuclear and Allied Radiations.

Medical Research Council (Great Britain), Committee on the Hazards to Man of Nuclear and Allied Radiations. 1956. *The hazards to man of nuclear and allied radiations.* London: Her Majesty's Stationery Office.

———. 1960. *The hazards to man of nuclear and allied radiations: A second report to the Medical Research Council.* London: Her Majesty's Stationery Office.

Medvedev, Zhores A. 1990. *The legacy of Chernobyl.* New York: W. W. Norton.

Michener, James Albert. 1954. *Sayonara.* New York: Random House.

Miller, R. W., and W. J. Blot. 1972. Small head size after in-utero exposure to atomic radiation. *Lancet* 2:784–87.

Miller, R. W., and J. J. Mulvihill. 1976. Small head size after atomic irradiation. *Teratology* 14:355–58.

Misao, T. 1961. Characteristics in abnormalities observed in atom-bombed survivors. *Journal of Radiation Research* 2:85–97.

Moloney, W. C., and M. A. Kastenbaum. 1955. Leukemogenic effects of ionizing radiation on atomic bomb survivors in Hiroshima City. *Science* 121 (25 February):308–9.

Mørch, E. H. 1941. Chondrodystrophic dwarfs in Denmark. *Opera Ex Domo Biologiae Hereditariae Humanae Universitatis Hafniensis.* 3:200.

Moriyama Isao. 1988. Secretariat 1948–84. Commemoration of the 40th anniversary of US-Japan joint studies of late A-bomb effects. *RERF Newsletter* 14 (1 March):43–45.

Morris, Edita. 1958. The Survivors of the Bombs. *New Statesman* 2 August.

Morton, Newton E. 1955. The inheritance of human birth weight. *Annals of Human Genetics* 20:125–34.

Muller, H. J. 1927. Artifical transmutation of the gene. *Science* 66:84–87.

———. 1948. Mutational prophylaxis. *Bulletin of the New York Academy of Medicine,* July, 447–69.

————. 1950a. Our load of mutations. *American Journal of Human Genetics* 2:111–176.

————. 1950b. Radiation damage to the genetic material: I. Effects manifested mainly in the descendants. *American Scientist,* January, 33–126.

————. 1954. The nature of the genetic effects produced by radiation. In *Radiation biology,* edited by Alexander Hollaender, 1:351–473. New York: McGraw-Hill.

————. 1956. Race poisoning by radiation. *Saturday Review,* 9 June, 9–39.

Mutagenesis experts brainstorm at RERF. 1991. *RERF Update* 3 (Winter 1991–92):1–2.

Myers, Greg. 1990. *Writing biology: Texts in the social construction of scientific knowledge.* Madison: University of Wisconsin Press.

Nadson, G. A., and G. S. Philippov. 1925. Influence des rayons X sur la sexualité et la formation des mutantes chez les champignons inférieures (Mucorinées). *Comptes Rendu Société Biologie* 93:473–75.

Nagai Takashi. 1958. *We of Nagasaki: The story of survivors in an atomic wasteland.* Translated by Ichiro Shirato and Herbert B. L. Silverman. New York: Meredith Press.

————. 1984. *The bells of Nagasaki.* Translated by William Johnston. Tokyo: Kodansha International.

Nagaoka Chizuno. 1973. A child whose mother was pregnant when A-bombed. In *Hiroshima in memoriam and today: Hiroshima as a testimony of peace for mankind,* edited by Takayama Hitoshi, 143–46. Hiroshima: Daigaku Letterpress.

National Academy of Sciences. 1989. *Questions and answers about the National Academy of Sciences, National Academy of Engineering, Institute of Medicine and National Research Council.* Washington, D.C.: National Academy of Sciences.

National Academy of Sciences, Committee on the Biological Effects of Atomic Radiation. 1960. *A report to the public on the biological effects of atomic radiation.* Washington D.C.: National Academy of Sciences.

National Academy of Sciences, National Research Council. 1956. *The biological effects of atomic radiation: A report to the public from a study.* Washington, D.C.: National Academy of Sciences, National Research Council.

National Commission on Excellence in Education. 1984. *A nation at risk: The full account.* Cambridge, Mass.: USA Research.

National Research Council, Committee on Atomic Casualties, Genetics Conference. 1947. Genetic effects of the atomic bombs in Hiroshima and Nagasaki. *Science* 106:331–33.

National Research Council of Japan, Special Committee for the Investigation of the Effects of the Atomic Bomb, Medical Section. 1953. *Medical report on atomic bomb effects.* Tokyo: Nankodo.

Neel, James V. 1937. Phenotypic variability in mutant characters of *Drosophila funebris. Journal of Experimental Zoology* 75:131–42.

————. 1940. The interrelations of temperature, body size and character expression in *Drosophila melanogaster. Genetics* 25:225–50.

————. 1941a. A relation between larval nutrition and the frequency of crossing over in the third chromosome of *Drosophila melanogaster. Genetics* 26:506–16.

———. 1941b. Studies on the interaction of mutations affecting the chaetae of *Drosophila melanogaster:* I. The interaction of hairy, polychaetoid and hairy wing. *Genetics* 26:52–68.

———. 1942a. The polymorph mutant of *Drosophila melanogaster. American Naturalist* 76:630–34.

———. 1942b. A study of a case of high mutation rate in *Drosophila melanogaster. Genetics* 27:519–36.

———. 1943a. Concerning the inheritance of red hair. *Journal of Heredity* 34:93–96.

———. 1943b. Studies on the interaction of mutations affecting the chaetae of *Drosophila melanogaster:* II. The relation of character expression to size in flies homozygous for polychaetoid, hairy, hairy wing and the combinations of these factors. *Genetics* 28:49–68.

———. 1947. The clinical detection of the genetic carriers of inherited disease. *Medicine* 26:115–53.

———. 1949. Genetic effects of the atomic bombs in Hiroshima and Nagasaki. *Science* 106:331–33.

———. 1957. Special problems inherent in the study of human genetics, with particular reference to the evaluation of radiation risks. *Proceedings of the National Academy of Sciences* 43:736–44.

———. 1958a. The delayed effects of ionizing radiation. *Journal of the American Medical Association* 166:908–16.

———. 1958b. A study of major congenital defects in Japanese infants. *American Journal of Human Genetics* 10:398–445.

———. 1963. *Changing perspectives on the genetic effects of radiation.* Springfield, Ill.: Charles C. Thomas.

———. 1967. The web of history and the web of life: Atomic bombs, inbreeding and Japanese genes. *Michgan Quarterly Review* 6:202–9.

———. 1970. Lessons from a "primitive" people: Do recent data concerning South American Indians have relevance to problems of highly civilized communities? *Science* 170:815–22.

———. 1973a. Our twenty-fifth: Speech presented at the twenty-fifth anniversary meeting of the American Society of Human Genetics at Atlanta, October 24–27, 1973. *American Journal of Human Genetics* 26:136–44.

———. 1973b. Social and scientific priorities in the use of genetic knowledge. In *Ethical issues in human genetics,* edited by Bruce Hilton, D. Callahan, M. Harris, P. Condliffe, and B. Berkley, 353–68. John E. Fogarty International Center for Advanced Study in the Health Sciences, Proceedings, no. 13. New York: Plenum Press.

———. 1981. Editorial: Genetic effects of atomic bombs. *Science* 213:1205.

———. 1985. The feasibility and urgency of monitoring human populations for the genetic effects of radiation: The Hiroshima-Nagasaki experience. In *Assessment of risk from low-level exposure to radiation and chemicals,* edited by A. D. Woodhead, C. J. Shelbarger, V. Pond, and A. Hollaender, 393–413. Basic Life Sciences, vol. 3. New York: Plenum Publishing.

Neel, James V., et al. 1980. See Neel, Satoh, Hamilton, Otake, Goriki, Kageoka, Fujita, Neriishi, and Asakawa, 1980.

Neel, James V., Gilbert Beebe, and Robert W. Miller. 1985. Delayed biomedical effects of the bombs. *Bulletin of the Atomic Scientists* 40 (August):72–75.

Neel, James V., and H. F. Falls. 1951. The rate of mutation of the gene responsible for retinoblastoma in man. *Science* 114:419–22.

Neel, James V., H. Kato, and W. J. Schull. 1974. Mortality in the children of atomic bomb survivors and controls. *Genetics* 76:311–26.

Neel, James V., C. Satoh, H. B. Hamilton, M. Otake, K. Goriki, T. Kageoka, M. Fujita, S. Neriishi, and J. Asakawa. 1980. Search for mutations affecting protein structure in children of atomic bomb survivors: Preliminary report. *Proceedings of the National Academy of Sciences* 77:4221–25.

Neel, James V., and William J. Schull. 1956a. *The effect of exposure to the atomic bombs on pregnancy termination in Hiroshima and Nagasaki.* In collaboration with R. C. Anderson, W. H. Borges, R. C. Brewer, S. Kitamura, M. Kodani, D. J. McDonald, N. W. Morton, M. Suzuki, K. Takeshima, W. J. Wedemeyer, J. W. Wood, S. W. Wright, and J. N. Yamazaki. National Academy of Sciences, National Research Council, publication no. 461. Washington, D.C.: National Academy of Sciences.

———. 1956b. Studies on the potential genetic effects of the atomic bombs. Proceedings, International Congress of Human Genetics, Copenhagen. *Acta Genetica* 6:183–96.

———. 1958. Radiation and the sex ratio in man. *Science* 128:343–48.

———. 1962. Genetic effects of the atomic bombs: Rejoinder to Dr. de Bellefeuille. *Acta Radiologica* 58:385–99.

Neel, James V., and William J. Schull, eds. 1991. *The children of atomic bomb survivors: A Genetic Study.* Washington, D.C.: National Academy of Sciences.

Neel, James V., William J. Schull, Duncan J. McDonald, Newton E. Morton, M. Kodani, K. Takeshima, R. C. Anderson, J. Wood, R. Brewer, S. Wright, J. Yamazaki, M. Suzuki, and S. Kitamura. 1953a. The effect of exposure of parents to the atomic bombs on the first generation offspring in Hiroshima and Nagasaki: Preliminary report. *Japanese Journal of Genetics* 28:211–18.

———. 1953b. The effect of exposure to the atomic bombs on pregnancy termination in Hiroshima and Nagasaki: Preliminary report. *Science* 118:537–41.

Neel, James V., William J. Schull, et al. 1989. The genetic effects of the atomic bombs: Problems in extrapolating from somatic cell findings to risk for children. In *Low dose radiation: Biological bases of risk assessment,* edited by K. F. Baverstock and J. W. Stather, 42–53. London: Taylor and Francis.

Neel, James V., and W. N. Valentine. 1944. Hematologic and genetic study of the transmission of thalassemia (Cooley's anemia, Mediterranean anemia). *Archives of Internal Medicine* 74:185–96.

Nishina Yoshio. 1947. A Japanese scientist describes the destruction of his cyclotrons. *Bulletin of the Atomic Scientists* 3 (June): 145, 167.

Nishiyama, H., R. E. Anderson, K. Ishimaru, Y. Li, and N. Okabe. 1973. The incidence of malignant lymphoma and multiple myeloma in Hiroshima and Nagasaki atomic bomb survivors, 1945–1965. *Cancer* 32:1301–9.

Ōe Kenzaburō. 1981. *Hiroshima notes*. Translated by Toshi Yonezawa and edited by David L. Swain. Tokyo: YMCA Press.

Osada Arata. 1951. *Genbaku no ko: Hiroshima no shonen to shojo no uttae* [Children of the A bomb: The testament of the boys and girls of Hiroshima]. Tokyo: Iwanami Shoten.

———. 1959. *Children of the A bomb: The testament of the boys and girls of Hiroshima*. Translated by Jean Dan and Ruth Sieben-Morgan. Tokyo: Uchida Rokakuho Publishing House.

Oughterson, Ashley W. 1955. *Statistical analysis of the medical effects of the atomic bomb*. Oak Ridge, Tenn.: Army Institute of Pathology, U.S. Atomic Energy Commission.

Oughterson, Ashley W., and Shields Warren, eds. 1956. *Medical effects of the atomic bomb in Japan*. National Energy Series, vol. 8, New York: McGraw-Hill.

Passin, Herbert. 1955. Japan and the H-bomb. *Bulletin of the Atomic Scientists* 9 (October):289–92.

Paul, Diane. 1987. Our load of mutations revisited. *Journal of the History of Biology* 20 (fall):3.

Perry, John Curtis. 1980. *Beneath the eagle's wings: Americans in occupied Japan*. New York: Dodd, Mead and Co.

Pharr, Susan J. 1987. The politics of women's rights. In *Democratizing Japan: The Allied occupation*, edited by Robert E. Ward and Yoshikazu Sakamoto. Honolulu: University of Hawaii Press.

Plummer, G. W. 1952. Anomalies occurring in children exposed in utero to the atomic bomb in Hiroshima. *Pediatrics* 10:687–93.

Proctor, Robert. 1988. *Racial hygiene: Medicine under the Nazis*. Cambridge, Mass.: Harvard University Press.

Radiation Effects Research Foundation. 1988. *A 40th year pictorial of the Atomic Bomb Casualty Commission and Radiation Effects Research Foundation*. Hiroshima: Radiation Effects Research Foundation.

Reingold, Nathan. 1985. Metro-Goldwyn Mayer meets the atom bomb. In *Expository science: Forms and functions of popularization*, edited by Terry Shinn and Richard Whitley. Boston: D. Reidel.

Reingold, Nathan, and Marc Rothenberg. 1987. *Scientific colonialism: A cross-cultural comparison*. Washington, D.C.: Smithsonian Institution.

Report on Hiroshima: Thousands of babies, no A-bomb effects. 1955. *U.S. News and World Report*, 8 April.

Reynolds, Earle L. 1961. *The forbidden voyage*. New York: D. McKay Co.

Rhodes, Richard. 1986. *The making of the atomic bomb*. New York: Simon and Schuster.

Roesch, W. C., ed. 1987. *U.S.-Japan joint reassessment of atomic bomb radiation do-*

simetry in Hiroshima and Nagasaki. Hiroshima: Radiation Effects Research Foundation.

Rothman, David J. 1991. *Strangers at the bedside: A history of how law and ethics transformed medical decision-making.* New York: Basic Books.

Russell, W. L. 1956. Comparison of X-ray-induced mutation rates in *Drosophila* and mice. *American Naturalist* 90:69–80.

Sasaki, M. S., and H. Miyata. 1968. Biological dosimetry in atomic bomb survivors. *Nature* 220:1189–93.

Schoenberger, Walter Smith. 1970. *Decision of destiny.* Athens, Ohio: Ohio University Press.

Schull, William J. 1953. The effect of Christianity on consanguinity in Nagasaki. *American Anthropologist* 55:74–88.

———. 1986. Scientist, journalist, orchidist: Will the real James V. Neel please stand up! In *Evolutionary perspectives and the new genetics: Proceedings of an international symposium honoring Dr. James V. Neel held in Ann Arbor Michigan, June 17–18, 1985,* edited by Henry Gershowitz, Donald L. Rucknagle, and Richard E. Tashian, 1–9. Progress in Clinical and Biological Research, vol. 218. New York: Alan R. Liss, 1986.

———. 1990. *Song among the ruins.* Cambridge, Mass.: Harvard University Press.

Schull, William J., and James V. Neel. 1958. Radiation and the sex ratio in man. *Science* 128:343–48.

———. 1959. Atomic bomb exposure and the pregnancies of biologically related parents: A prospective study of the genetic effects of ionizing radiation in man. *American Journal of Public Health* 49:1621–29.

———. 1962. Letter to the editor: Maternal radiation and mongolism. *Lancet* 1:537–38.

———. 1965. *The effects of inbreeding on Japanese children.* New York: Harper and Row.

———. 1972. The effects of parental consanguinity and inbreeding in Hirado, Japan: V. Summary and interpretation. *American Journal of Human Genetics* 24:425–53.

Schull, William J., James V. Neel, and A. Hashizume. 1966. Some further observations on the sex ratio among infants born to survivors of the atomic bombings of Hiroshima and Nagasaki. *American Journal of Human Genetics* 18:328–38.

———. 1981. Some further observations on the sex ratio among infants born to survivors of the atomic bombs: A reappraisal. *Science* 213:1220–27.

Searle, A. G. 1974. Mutation induction in mice. In *Advances in radiation biology,* edited by J. T. Lett, H. I. Adler, and A. Zelle, 4:131–207. New York: Academic Press.

Secord, James A. 1982. King of Siluria: Roderick Murchison and the imperial theme in nineteenth century British geology. *Victorian Studies* 25:413–42.

Seigel, D. W. 1966. Frequency of live births among survivors of Hiroshima and Nagasaki atomic bombs. *Radiation Research* 28:278–88.

Selden, Kyoko, and Mark Selden. 1989. *The atomic bomb: Voices from Hiroshima and Nagasaki.* Armonk, N.Y.: M. E. Sharpe.

Shapiro, Jacob. 1990. *Radiation protection: A guide for scientists and physicians.* Cambridge, Mass.: Harvard University Press.

Sherwin, Martin J. 1975. *A world destroyed: The atomic bomb and the grand alliance.* New York: Knopf.

Shiva, V., and J. Bandyopadhay. 1980. The large and fragile community of scientists in India. *Minerva* 18:575–94.

Shohno Naomi and Sakuma Kiyoshi. 1958. The fundamental examination of the amount of radiation received and its correlation to acute symptoms. In *Physical and medical effects of the atomic bombs in Hiroshima,* by the Hiroshima Research Group of Atomic Bomb Casualty, 1–27. Tokyo: Maruzen Co.

Simons, Marlise. 1993. Soviet atom test used thousands as guinea pigs, archives show. *New York Times,* 7 November.

Snell, F. M., J. V. Neel, and K. Ishibashi. 1949. Hematologic studies in Hiroshima and a control city two years after the atomic bombing. *Archives of Internal Medicine* 84:569–604.

Snell, G. D. 1935. The induction by X-rays of hereditary changes in mice. *Genetics* 20:545–67.

Steinberg, Arthur G. 1957. *Science,* 1 November, 931–32.

Stern, Curt. 1950. *Principles of human genetics.* San Francisco: W. H. Freeman and Co.

———. 1960. *Principles of human genetics.* 2d ed. San Francisco: W. H. Freeman and Co.

Stewart, Alice. 1971. Low dose radiation cancers in man. *Advances in Cancer Research* 14:359–90.

———. 1982. Delayed effects of A-bomb radiation: A review of the recent mortality rates and risk estimates for five year survivors. *Journal of Epidemiology and Community Health* 36:80–86.

———. 1985. Detection of late effects of ionizing radiation: Why deaths of survivors are so misleading. *International Journal of Epidemiology* 14:1, 52–56.

Stewart, Alice, and G. W. Kneale. 1988. Late effects of A-bomb radiation: Risk problems unrelated to the new dosimetry. *Health Physics* 54 (May): 567–69.

Strobel, D., and F. Vogel. 1958. Ein Statscher Gesichtspunkt fur das Planen von Untersuchungen uber Anderungen der Mutationsrate bein Menschen. *Acta Genetica* 8:274.

Takayama Hitoshi, ed. 1973. *Hiroshima in memoriam and today: Hiroshima as a testimony of peace for mankind.* Hiroshima: Daigaku Letterpress.

Tale of two cities. 1962. *Time,* 18 May, 22.

Tarnower, Herman, and Sam Sinclair Baker. 1978. *The complete Scarsdale medical diet plus Dr. Tarnower's lifetime keep-slim program.* New York: Rawson, Wade Publishers.

Timofeef-Ressovsky, Nikolai V. 1934. The experimental production of mutations. *Biological Reviews* 9:411–57.

Totten, George O., and T. Kawakami. 1964. Gensuikyō and the peace movement in Japan. *Asian Survey* 4:833–41.

Trumbull, Robert. 1957. *Nine who survived Hiroshima and Nagasaki: Personal experiences of nine men who lived through both atomic bombings.* New York: E. P. Dutton and Co.

Turpin, R., J. Lejeune, and M. O. Rethore. 1956. Etude de la descendants de sujets traites par radiotherapie pelvienne. *Acta Genetica et Statistica Medica* 6:204–16.

United Nations. See United Nations, Scientific Committee on the Effects of Atomic Radiation.

United Nations, Scientific Committee on the Effects of Atomic Radiation. 1958. *Report of the United Nations Scientific Committee on the Effects of Atomic Radiation.* Official Records, United Nations General Assembly, Supplement no. 17. New York: United Nations.

———. 1977. *Sources and effects of ionizing radiation: 1977 report to the General Assembly, with annexes.* New York: United Nations.

United States Atomic Energy Commission. 1960. *Genetics research program of the division of biology and medicine.* Washington, D.C.

United States Strategic Bombing Survey. 1946. *The effects of atomic bombs on Hiroshima and Nagasaki.* Washington, D.C.: U.S. Government Printing Office.

———. 1947. *The effects of strategic bombing on Japanese morale.* United States Strategic Bombing Survey, Reports, Pacific War, no. 14. Washington, D.C.: U.S. Government Printing Office.

———. 1973. *The effects of the atomic bombs on Hiroshima and Nagasaki.* Edited by William Gannon. Sante Fe, N.M.: Gannon.

United States Strategic Bombing Survey, Morale Division, Medical Branch. 1945. *The effect of bombing on health and medical care in Germany.* Washington, D.C.: War Department, 30 October.

Valentine, W. N., and J. V. Neel. 1945. The frequency of thalassemia. *American Journal of Medical Science* 209:568–72.

Vining, Elizabeth Gray. 1952. *Windows for the crown prince.* Philadelphia: Lippincott.

Wallace, Bruce, and Theodosius Dobzhansky. 1959. *Radiation, genes and man.* New York: Holt, Rinehart and Winston.

Ward, Robert E., and Yoshikazu Sakamoto, eds. 1987. *Democratizing Japan: The Allied occupation.* Honolulu: University of Hawaii Press.

Warren, Shields. 1942a. Blood findings in cyclotron workers. *Radiology* 39:194–99.

———. 1942b. Effects of radiation on normal tissues. *Archives of Pathology* 34:443–50.

Warren, Stafford L. 1966. *The role of radiation in the development of the atomic bomb in radiology in World War II.* Washington: U.S. Army, Medical Department.

Weart, Spencer R. 1988. *Nuclear fear: A history of images.* Cambridge, Mass.: Harvard University Press.

Weiner, Charles. 1978. Retroactive saber rattling? *Bulletin of the Atomic Scientists* 34 (April):10–12.

Wesley, J. P. 1960. Background radiation as the cause of fatal congenital anomalies. *International Journal of Radiation Biology* 2:97.

Whitney, Courtney. 1956. *MacArthur: His rendezvous with history.* New York: Alfred A. Knopf.

Whittemore, Gilbert F. 1986. The National Committee on Radiation Protection, 1928–1960: From professional guidelines to government regulation. Ph.D. dissertation, Harvard University.

Wildes, Harry Emerson. 1954. *Typhoon in Tokyo: The Occupation and its aftermath.* New York: Macmillan Co.

Willoughby, Charles A., and John Chamberlain. 1954. *MacArthur, 1941–1951.* New York: McGraw-Hill.

Wilson, Robert R. 1956. Nuclear radiation at Hiroshima and Nagasaki. *Radiation Research* 4:349.

Wood, J. W., K. G. Johnson, and Y. Omori. 1967. In utero exposure to the atomic bomb: An evaluation of head size and mental retardation twenty years later. *Pediatrics* 39:385–92.

World Health Organization. 1957. *Effect of radiation on human heredity: A report of a study group convened by WHO, together with papers presented by various members of the group.* Geneva: World Health Organization.

Wyden, Peter. 1984. *Day one: Before Hiroshima and after.* New York: Simon and Schuster.

Yale and the ABCC. 1968. *Alumnae Bulletin,* spring, 7–10. New Haven, Conn.: Yale University.

Yavenditti, Michael J. 1974. John Hersey and the American conscience: The reception of *Hiroshima. Pacific Historical Review,* October, 24–29.

Yoshimoto, Y., J. V. Neel, W. J. Schull, H. Kato, M. Soda, R. Eto, and K. Mabuchi. 1990. Malignant tumors during the first two decades of life in the offspring of atomic bomb survivors. *American Journal of Human Genetics* 46:1041–52.

Ziemke, Earl F. 1975. *The U.S. Army in the occupation of Germany, 1944–1946.* Washington, D.C.: U.S. Army, Center of Military History.

———. 1984. Improvising in post-war Germany. In *Americans as proconsuls: United States military government in Germany and Japan, 1944–1952,* edited by Robert Wolfe, 52–66. Carbondale: Southern Illinois University Press.

JAPANESE LANGUAGE MATERIALS

Anchi hyūman: Dokyumento nihonjin 8 [Anti-human: Documents of the Japanese people 8]. 1969. Tanigama Ken'ichi, Surumi Shunsuke, and Murakami Ichirō, eds. Tokyo: Gakugie Shorin. Translated into English by Jaylene Masaoka, Wilmington College Peace Resource Center, Wilmington, Ohio.

Gembaku, Gohyakunin no shōgen: Hibakusha tsuiseki chōsa repōto [Five hundred a-bomb testimonies: Report of a hibakusha survey]. 1967. Tokyo: Asahi Shimbum. Survivors were asked their opinions of the Atomic Bomb Casualty Commission. Available in Japanese at the Wilmington College Peace Resource Center, Wilmington, Ohio.

Gensuibaku higai hakusho: Kakusareta shinjitsu [White paper on hydrogen and atomic bomb victims: The hidden truth]. 1961. Gensuibaku kinshi nihon kyōgikai senmon iinkai, eds. [Technical commitee of the Japan Council Against Atomic and Hydrogen Bombs]. Tokyo: Nihon Hyōron Shinsa. An unpublished translation is avaiable at the Wilmington College Peace Resource Center, Wilmington, Ohio.

Henken to Sabetsu: Hiroshima soshite hibaku chōsenjin [Prejudice and discrimination: Hiroshima and Korean Hibakusha]. 1971. Hiraoka Takashi. Tokyo: Miraisha. Compilation of articles published 1966–72 dealing with the prejudice and discrimination to which *hibakusha* were subjected. Available in Japanese at the Wilmington College Peace Resource Center, Wilmington, Ohio.

Hibaku nisei: Sono katararenakatta hibi to asu [Hibaku nisei: Their untold past and their futures]. 1972. Hiroshima kishadan hibaku nisei kankō iinkai, ed. Tokyo: jiji Tshūshin Sha. This text explores the possibility of genetic effects in the children of survivors, presenting cases of leukemia in some survivor children and arguing that the children of survivors should qualify for medical benefits as victims of the atomic bombings. Available in Japanese at the Wilmington College Peace Resource Center, Wilmington, Ohio.

Kagaisha e no ikari: ABCC wa nani o shita ka [Anger toward the aggressor: How the ABCC treated us]. 1966. Hiroshima Gensuibaku Kinshi Hiroshima-shi kyōgikai. This book contains accounts by survivors expressing anger at the ABCC, at being examined by ABCC personnel, and at being treated like "laboratory animals." One of the accounts is by a non-survivor who served in the control group and expressed similar angry feelings toward the ABCC. Available in Japanese at the Wilmington College Peace Resource Center, Wilmington, Ohio.

Kajiyama Toshiyuki. *Jikken Toshi* [Experimental City]. First published in the Japanese magazine *L'espoir* (1954). English translation in the papers of Robert Jay Lifton, New York City Public Library.

Index

Lilienthal, David, 105
Lucky Dragon: and ABCC no-treatment policy, 124–25, 133–34, 156; Japanese fishermen aboard, 5–6
Lyon, Mary, 172n

Ma No Isan (novel), 150
MacArthur, Douglas, 20–23, 84–85
Machle, Willard, 109–14, 110n
Macht, Stanley, 235–37
Macklin, Madge Thurlow, 78n
Maki Hiroshi, 94, 139, 146, 150, 180
Manhattan Project: estimate of radiation released by bombs, 28n; funding of radiation research, 105–6; medical team of, 22; and mouse work, 67; Neel's interest in, 69; and radiation sickness, 11–12
Marshall Islanders, 157, 217–18
Martin, Robert P., 148
Masaichi Miyamoto, 126
Masaoka Akira, 130, 162
Mather, Kenneth, 185
Matsubayahsi Ikuzo: Hiroshima survey of, 71–73; payments to midwives, 90; circulated questionnaires, 44; Hiroshima luncheon (1946), 43
Matsumoto, Scott, 154–56
Mavor, James Watt, 62
McDonald, Duncan: and early spontaneous abortion study, 183–85; and minimizing radiation effects, 214–15; and 1952–53 analysis, 193–215; and midwives payments, 91; and 1953 announcement of results, 220–32
McIntyre, Ross, 22
measures of radiation, 66n
medical care, and ABCC, 117–42
medical licensing, and ABCC no-treatment policy, 134–35
Mercer, George I., 135–36
Michener, James, 83
microcephaly, 100, 126n
midwives: and birth questionnaire, 87–88, 92; concessions to, 98; conflicts provoked by cooperation with ABCC, 91–94; and crematorium, 94–95; education and training of, 89–90; and Matsubayashi study, 90; meetings with, 42, 88–89; payments to, 73, 88, 90, 91; relationship to families, 88, 90; role in genetics study, 85–101; weighing of newborns, 193
minor malformation, 186–92
Moloney, William, 126
Moore, Felix E., 245
Morito Tatsuo, 161
Morton, John, 125, 157–58
Morton, Newton, 105; and analysis (1952–53), 193–215; and early spontaneous abortion study, 183; and publication of genetics results, 227
mouse data: as comparable to human, 224–25; and extrapolation, 115, 191; and genetic effects, 63–65
Muller, H. J.: doubts about the ABCC genetics study, 73–74; early work, 62; and fertility effects, 176; at the 1947 Genetics Conference, 70–79; and mutation rates in man, 74; and sex ratio, 172; mentioned, 105, 145, 148, 238
Mumford, Lewis, 137
mutations: analysis of, 169–92; invisible, 73–74, 77

Nagai Isamu, 139
Nagai Takashi, 48
Nagasaki: August 1945 bombing, 3–6, 28; foreign populations in, 234; population changes, 207
Nagasaki Midwives Association, 89–90
Nagasaki University, 152
Napier, John T., 135–36, 136n
National Academy of Science: reports to the public, 218; and sponsorship of ABCC, 32–33, 249; mentioned, 107, 194
National Council on Radiation Protection, 63, 245n

Wright, Stanley, 187
Wyden, Peter, 11–12

X rays, 61–62
Yale University, 247–48

Yamamoto Hisao, 43
Yamamoto Setsuko, 91, 94–95
Yoshiko Yamagushi, 145

Zelle, Max, 114–15
Zirkle, Raymond, 76